Research in Law Enforcement Selection

Michael G. Aamodt, PhD
Radford University

BrownWalker Press
Boca Raton • 2004

Research in Law Enforcement Selection

BrownWalker Press
Boca Raton , Florida
USA • 2004

ISBN: 1-58112-428-7

BrownWalker.com

Contents

Preface

The purpose of this book is to provide a reference source for individuals interested in law enforcement selection. The chapters contain meta-analyses (statistical reviews of the literature) investigating the validity of methods used to predict police performance. These methods include education requirements, cognitive ability, background variables (e.g., military experience, arrest record, discipline problems at work), personality inventories, interest inventories, physical agility tests, assessment centers, and interviews.

The first chapter in the book is a short primer on meta-analysis that informs the reader about the purpose of meta-analysis and how to interpret the meta-analysis tables contained in the book. Readers not familiar with meta-analysis should feel comfortable with the topic after reading Chapter 1. Readers already familiar with meta-analysis can probably skip this chapter.

Chapter 2 describes the methods used to conduct the meta-analyses for this project. Chapters 3-11 list the meta-analysis results for the various predictors of police performance. Chapter 12 describes a meta-analysis of the relationships among criteria (e.g., performance ratings, discipline problems, commendations), Chapter 13 describes a meta-analysis of the relationships among selection methods, and Chapter 14 describes a meta-analysis of the relationship between the various criteria and sex, race, age, and tenure. Chapter 15 summarizes the previous chapters and identifies future research needs.

The chapters were intentionally kept short. Each chapter includes a brief discussion of the topic, tables listing the meta-analysis results, and a brief discussion of the results. The beauty of meta-analysis is that the results tables typically tell the whole story of how well a selection method predicts performance and little additional discussion is usually needed.

For the reader who would like more information on a particular study, a PDF file is available containing over 400 pages of statistical summaries. These summaries contain the information used to conduct the meta-analyses described in chapters 3-14 and information that can be used for future meta-analyses. The aim of these summaries is to include enough information about a study that the reader will not need to consult the original source—an advantage when many of the original sources such as theses and dissertations can be difficult to obtain. Each summary contains complete citation information as well as information about the sample, the predictors and criteria used in the study, and the essential findings of the study. The web address for this file is www.radford.edu/~maamodt/riles.htm.

If I have done my job properly, with a few exceptions of dissertations that could not be obtained, summaries of all journal articles, theses, and dissertations relevant to this topic from 1970 – 2003 are in this book. To find studies relevant to this project, studies older than 1970 and more recent than September, 2003 were included when found but inclusion outside of the years 1970-2003 would not be considered exhaustive.

Acknowledgements

I would like to thank my wife Bobbie for her support in working on this 10-year project. My guess is if she never hears the words "meta-analysis" or "police selection" again she wouldn't complain. I am deeply indebted to JoAnne Brewster at James Madison University, Mark Foster at the University of Georgia Institute for Government, Wayman Mullins at Texas State University, Mark Nagy at Xavier University, Bobbie Raynes at New River Community College, and Michael Surrette at Springfield College for reading drafts of this book and providing useful feedback. They spent more hours reading drafts than I would ever have expected, and for that I am greatly appreciative.

I would like to also thank Bud Bennett and his interlibrary loan staff at Radford University. Bud's hard work and patience in getting other libraries to send dissertations, theses, and articles is much appreciated and this project could not have been completed without his help. Finally I would like to thank my colleagues in the Society for Police and Criminal Psychology for their support and encouragement throughout this project and to Radford University for granting me a one-semester sabbatical to finish the book.

Chapter 1
Introduction to Meta-Analysis

In the old days (prior to 1980) research was reviewed by reading all of the articles on a topic and then drawing a conclusion. For example, suppose that a personnel analyst was asked to review the literature to see if education was related to police performance. The analyst would find every article on the topic, perhaps count the articles showing significant results, and then reach a conclusion such as "given that eight articles showed significant results and nine did not, we must conclude that education is not related to police performance."

Unfortunately, there are three common situations in which such a conclusion might be inaccurate: small but consistent relationships, moderate relationships but small sample sizes, and large differences in sample sizes across studies.

Small but Consistent Correlations

Suppose that you find four studies investigating the relationship between education and police performance, but none of the four studies reported a significant relationship between the two variables. With a traditional review, you would probably conclude that education is not a significant predictor of police performance. However, it might be that the actual relationship between education and performance is relatively small, and a large number of subjects would have been needed in each study to detect this small relationship. Take for example the studies shown in Table 1.1. You have four studies, each with samples of 50 officers. The correlations between education level and performance in the four studies are .20, .17, .19, and .16. Though the size of the coefficients is consistent across the four studies, none of the correlations by itself is statistically significant due to the combination of small correlations and small sample sizes in each study. If the four studies are combined in a meta-analysis however, we find that the average correlation is .18 and with a sample size of 200, the correlation would be statistically significant.

Table 1.1
Example of a small but consistent relationship

Study	Correlation	Sample Size
Hill (1991)	.20	50
Renko (1992)	.17	50
Bates (1993)	.19	50
Coffey (1994)	.16	50

Moderate Relationships and Small Sample Sizes

A second situation in which traditional literature reviews often draw incorrect conclusions occurs when the correlations in the previous studies are moderate or high, but the sample sizes were too low for the relationship to be statistically significant. Take for example the four studies shown in Table 1.2. Each of the correlations is at what we would consider a high level, yet the correlations would not be statistically significant due to the small sample sizes in each study. If we combined the four studies however, we would get an average correlation of .41—with a total sample size of 80, this would be statistically significant.

Table 1.2
Example of large correlations but small sample sizes

Study	Correlation	Sample Size
Spencer (1998)	.43	20
Magnum (1997)	.38	20
Rockford (1992)	.45	20
Mannix (1998)	.39	20

Large Differences in Sample Sizes Across Studies

Another reason that we might incorrectly conclude that education does not predict performance is that differences in correlations across studies may be due to large differences in sample sizes. As you can see in the example shown in Table 1.3, the reason our traditional review would find mixed results is that the two studies showing a low correlation between education and performance had very small sample sizes. Thus, what seem to be huge differences in validity are actually differences due to *sampling error* caused by small sample sizes.

Table 1.3
Example of inconsistent sample sizes

Study	Correlation	Sample Size
Sullivan (1998)	.35	400
Davis (1997)	.05	20
Boscorelli (1992)	.40	290
Yokus (1988)	.10	20

To better understand sampling error, imagine that you have a bowl containing three red balls, three white balls, and three blue balls. You are then asked to close your eyes and pick three balls from the bowl. Because there are equal numbers of red, white, and blue balls in the bowl, you would expect to draw one of each color. However, in any given draw from the bowl, it is unlikely that you will get one of each color. If you have no life and draw three balls at a time for ten hours, you might get three red balls on some draws, three white balls on other draws, and three blue balls on other draws. Thus, even though we know there are an equal number of each color of ball, any one draw may or may not represent what we know is "the truth." However, over the 10 hours you are drawing balls, the most common draw will be one of each color—a finding consistent with what we know is in the bowl.

The same is true in research. Suppose we know that the true correlation between education level and performance is .20.

A study at one agency might yield a correlation of .10, another agency might report a correlation of .50, and yet another agency might report a correlation of .30. If all three studies had small samples, the differences among the studies and differences from the "truth" might be due purely to sampling error. This is where meta-analysis saves the day.

Meta-analysis is a statistical method for combining research results. Since the first meta-analysis was published by Gene Glass in 1976, the number of published meta-analyses has increased tremendously and the methodology has become increasingly complex. The most influential meta-analysts are Frank Schmidt and the late John Hunter, and almost every meta-analysis uses the methods suggested in their 1990 book *Methods of Meta-Analysis* and clarified in the book *Conducting Meta-Analysis Using SAS* by Winfred Arthur, Winston Bennett, and Allen Huffcutt (2001).

Though meta-analyses will vary somewhat in their methods and their purpose, most meta-analyses involving personnel selection issues such as those discussed in this book try to answer three questions:

1) What is the mean validity coefficient found in the literature for a given selection method (e.g., education, interviews, assessment centers, cognitive ability)?

2) If we had a perfect measure of the predictor (e.g., intelligence, computer knowledge), a perfect measure of performance, and no restriction in range, what would be the "true correlation" between our selection method and performance?

3) Can we generalize the meta-analysis results to every law enforcement agency (validity generalization), or is our selection method a better predictor of performance in some situations than in others (e.g., large vs. small departments, police departments vs. sheriff's offices)?

Conducting a Meta-Analysis

Finding Studies

The first step in a meta-analysis is to locate studies on the topic of interest. It is common to use both an "active search" and a "passive search." An active search tries to identify every research study within a given parameter. For example, a meta-analyst might concentrate her active search on journal articles and dissertations published between 1970 and 2001 and referenced in one of three computerized literature databases (PsycInfo, Infotrac, Dissertation Abstracts International) or referenced in an article found during the computer search. A passive search might include queries to professionals known to be experts in the area, papers presented at conferences, or technical reports known to the author.

The major difference between an active and passive search is that the goal of an active search is to include *every* relevant study within the given parameters, whereas the goal of the passive search is to find relevant research without any thought that every study on the topic was found. Though this may not seem like much of a difference, it is. These days, there are so many potential sources for research—thousands of journals, conference presentations, theses, dissertations, technical reports, and unpublished research articles—that relevant studies are going to be missed. Thus the credibility of a meta-analysis hinges on the scope and inclusion accuracy of its active search.

Choosing Studies to Include in the Meta-Analysis

Once the relevant studies on a topic have been located, the next step is to determine which of these studies will be included in the meta-analysis. To be included in a meta-analysis, an article must report the results of an empirical investigation and must include a correlation coefficient, another statistic (e.g., F, t, chi-square) that can be converted into a correlation coefficient, or tabular data that can be entered into the computer to yield a correlation coefficient. Articles that report results without the above statistics (e.g., "We found a significant relationship between education and academy performance" or "We didn't see any real differences between our educated and uneducated officers") cannot be included in a meta-analysis.

Often, meta-analysts will have other rules about including studies. For example, in a meta-analysis on employee-wellness programs, the researcher's decision to include only studies using both pre- and post-measures of absenteeism as well as experimental and control groups resulted in only three usable studies.

Converting Research Findings to Correlations

Once research articles have been located and the decision made as to which articles to include, statistical results (e.g., F, t, Chi-square) that need to be converted into correlation coefficients are converted using the formulas provided in Arthur et al. (2001). In some cases, raw data or data listed in tables can be entered into a statistical program (e.g., SAS, SPSS) to directly compute a correlation coefficient.

Cumulating Validity Coefficients

As shown in Table 1.4, after the individual correlation coefficients have been computed, the validity coefficient for each study is weighted by the size of the sample, summed, and then divided by the total sample size to arrive at a mean validity coefficient. This procedure ensures that larger studies—presumed to be more accurate—carry more weight than smaller studies. For example, in Table 1.4, the .23 correlation reported by Briscoe is multiplied by the sample size of 150 to get 34.5. This procedure is then done for each of the studies.

Table 1.4
Example of cumulating validity coefficients

Study	Correlation	Sample Size	Correlation x Sample Size
Briscoe (1997)	.23	150	34.5
Green (1974)	.10	100	10.0
Curtis (1982)	.42	50	21.0
Logan (1991)	.27	300	81.0
Ceretta (1995)	.01	20	0.2
Greevy (1989)	.29	200	58.0
TOTAL		820	204.8

Weighted Average = 204.8 ÷ 820 = .25

Correcting for Artifacts

When conducting a meta-analysis, it is desirable to adjust correlation coefficients to correct for error associated with such study artifacts as sampling error, measurement error (predictor unreliability, criterion unreliability), dichotomization, and restriction of range (see Hunter & Schmidt, 1990 for a thorough discussion). These adjustments answer the second question of, "If we had perfect measures of the construct, a perfect measure of performance, and no restriction in range, what would be the "true correlation" between our construct and performance?"

These adjustments can be made in one of two ways. The most desirable way is to correct the validity coefficient from each study on the basis of the predictor reliability, criterion reliability, and restriction of range associated with that particular study. A simple example of this process is shown in Table 1.5. To correct the validity coefficients in each study for test unreliability, the validity coefficient is divided by the square root of the reliability coefficient. In the Baker (1985) study, the reliability of the test was .92, the square root of .92 is .96, and the corrected validity coefficient is .30 ÷ .96 = .31.

Table 1.5
Correcting correlations for test unreliability

Study	Validity	Test Reliability	Square-Root	Corrected Validity
Baker (1985)	.30	.92	.96	.31
Poncherella (1990)	.23	.80	.89	.26
Baricza (1995)	.25	.65	.81	.31

When the necessary information is not available for each study, the mean validity coefficient is corrected rather than each individual coefficient. This is the most common practice. The numbers used to make these corrections come either from the average of information found in the studies that provided reliability or range restriction information or from other meta-analyses. For example, an estimate of the reliability of supervisor ratings of overall performance (r = .52) can be borrowed from the meta-analysis on rating reliability by Viswesvaran, Ones, and Schmidt (1996).

Searching for Moderators

Being able to generalize meta-analysis findings across all similar organizations and settings (validity generalization) is an important goal of any meta-analysis. It is standard practice in meta-analysis to generalize results when at least 75% of the observed variability in validity coefficients can be attributed to sampling and measurement error. When less than 75% can be attributed to sampling and measurement error, a search is conducted to find variables that might moderate the size of the validity coefficient. For example, education might predict performance better in larger police departments than in smaller ones.

The idea behind this 75% rule is that due to sampling and measurement error, we expect correlations to differ from study to study. The question is, are the differences we observe just sampling and measurement error, or do they represent real differences in studies? That is, is the difference between the correlation of .30 found in one study and the correlation of .20 found in another study due to sampling error, or is the difference due to one study being conducted in an urban police department and the other being conducted in a rural department?

To answer this question there are formulas that tell us how much variability in studies we have in our meta-analysis, and how much of that variability would be expected due to sampling error and how much due to measurement error.

Understanding Meta-Analysis Results

Now that you have an idea about how a meta-analysis is conducted, let's talk about how to understand the meta-analysis results you will find in this book. In Table 1.6 you will find the partial results of a meta-analysis conducted on the relationship between cognitive ability and police performance. The numbers in the table represent the validity of cognitive ability in predicting academy grades and supervisor ratings of performance as a police officer.

Number of Studies and Sample Size

The "K" column indicates the number of studies included in the meta-analysis and the "N" column indicates the number of total subjects in the studies. There is not a magical number of studies we look for but a meta-analysis with 20 studies is clearly more useful than one with 5.

Mean Observed Validity Coefficient

The "*r*" column represents the mean validity coefficient across all studies (weighted by the size of the sample). This coefficient answers our question about the typical validity coefficient found in validation studies on the topic of cognitive ability and police performance. On the basis of our meta-analysis, we would conclude that the validity of cognitive ability in predicting academy grades is .41 and the validity of cognitive ability in predicting supervisor ratings of on-the-job performance is .16.

Confidence Intervals

To determine if our observed validity coefficient is "statistically significant," we look at the next two columns that represent the lower and upper limits to our 95% confidence interval. If the interval includes zero, we cannot say that our mean validity coefficient is significant. From the figures in Table 1.6, we would conclude that cognitive ability is a significant predictor of grades in the academy (our confidence interval is .33 - .48) and performance as a police officer (our confidence interval is .12 - .20).

Table 1.6
Sample meta-analysis results for cognitive ability

Criterion	K	N	r	95% Confidence Interval		ρ	90% Credibility Interval		Var	Q_w
				Lower	Upper		Lower	Upper		
Academy Grades	61	14,437	.41	.33	.48	.62	.44	.81	78%	77.82
Supervisor Ratings	61	16,231	.16	.12	.20	.27	.14	.40	80%	76.40

K=number of studies, N=sample size, r = mean correlation, ρ = mean correlation corrected for range restriction, criterion unreliability, and predictor reliability, VAR = percentage of variance explained by sampling error and study artifacts, Q_w = the within group heterogeneity

Using confidence intervals we can communicate our findings with a sentence such as, "Though our best estimate of the validity of cognitive ability in predicting academy performance is .41, we are 95% confident that the validity is no lower than .33 and no higher than .48." It is important to note that some meta-analyses use 80%, 85%, or 90% confidence intervals. The choice of confidence interval levels is a reflection of how conservative a meta-analyst wants to be: The more cautious one wants to be in interpreting the meta-analysis results, the higher the confidence interval used.

Corrections for Artifacts

The column labeled ρ (rho) represents our mean validity coefficient corrected for criterion unreliability, predictor unreliability, and range restriction. This coefficient represents what the "true validity" of cognitive ability would be if we had a perfectly reliable measure of cognitive ability, a perfectly reliable measure of academy grades and supervisor ratings of performance, and no range restriction. Notice how our observed correlations of .41 and .16 increase to .62 and .27 after being corrected for study artifacts. When encountering ρ, it is important to consider how many of the artifacts were included in the corrected correlation. That is, two meta-analyses on the same topic might yield different results if one meta-analysis corrected for all three artifacts while another corrected only for criterion unreliability.

Credibility Interval

Credibility intervals are used to determine if the corrected correlation coefficient (ρ) is statistically significant and if there are moderators present. Whereas a standard deviation is used to compute a confidence interval, the standard error is used to compute a credibility interval. As with confidence intervals, if a credibility interval includes zero, the corrected correlation coefficient is not statistically significant. If a credibility interval contains zero or is large, the conclusion to be drawn is that the corrected validity coefficient cannot be generalized and that moderators are operating (Arthur, Bennett, & Huffcutt, 2001). When reading a meta-analysis table, caution must be taken as the abbreviation CI is often used both for confidence and credibility intervals.

Percentage of Variance Due to Sampling Error and Study Artifacts

The next column in a meta-analysis table represents the percentage of observed variance that is due to sampling error and study artifacts. Notice that for grades and performance, these percentages are 78% and 80% respectively. Because the percentage is greater than 75, we can generalize our findings and do not need to search for moderators. Such a finding is desired, but is unusual. More typical is the meta-analysis results shown in Table 1.7. These results are from the excellent meta-analysis of the relationship between grades in school and work performance that was conducted by Roth, BeVier, Switzer, and Schippmann (1996).

Roth and his colleagues found that only 54% of the observed variance in correlations would have been expected by sampling error and study artifacts. Because of this, they were forced to search for moderators. They hypothesized that the level of education where the grades were earned (undergraduate, masters, or doctoral program) might moderate the validity of how well grades predicted work performance. As you can see from the table, the validity of grades in master's degree programs was higher than in doctoral programs, and sampling error and study artifacts explained 100% of the variability across studies for these two levels. However, sampling error and study artifacts accounted for only 66% of the observed variance in correlations for grades earned at the bachelor's level. So, the researchers further broke the bachelor's level grades down by the years since graduation.

Rather than using the 75% rule, some meta-analyses use a test of statistical significance. The results of these tests are reported as a Q_w or H_w statistic. If this statistic is significant, then a search for moderators must be made. If the statistic is not significant, we can generalize our findings. As shown back in

Table 1.6, the Q_w statistic was not significant for either academy grades or supervisor ratings of performance. This lack of significance is consistence with the fact that sampling error and study artifacts accounted for at least 75% of the observed variance.

A good example of the use of this statistic can be found in a meta-analysis of the effect of flextime and compressed workweeks on work-related behavior (Baltes, Briggs, Huff, Wright, & Neuman, 1999). As you can see in Table 1.8, the asterisks in the final column indicate a significant Q_w, forcing a search for moderators. Note that this meta-analysis used the *d* statistic rather than an *r* (correlation) as the effect size. In this example, a *d* of .30 is equivalent to an *r* of .15.

Now that you have a basic understanding of meta-analysis, the next chapter will describe the specific details of how the meta-analyses were conducted for this book.

Table 1.7
Meta-analysis of grades and work performance

Criterion	K	N	r	r_{cr}	$r_{cr,rr}$	$r_{cr,rr,pr}$	80% C.I.	SE%
Overall	71	13,984	.16	.23	.32	.35	.30-.41	54
Education Level								
B.A.	49	9,458	.16	.23	.33	.36	.30-.42	66
M.A.	4	446	.23	.33	.46	.50	.31-.56	100
Ph.D./M.D.	6	1,755	.07	.10	.14	.15	.08-.25	100
Years since graduation								
1 year	13	1,288	.23	.32	.45	.49	.40-.62	89
2-5 years	11	1,562	.15	.21	.30	.33	.23-.48	80
6+ years	4	866	.05	.08	.11	.12	.00-.41	59

Table 1.8
Meta-analysis of flextime and compressed work weeks

Variable	K	N	*d*	95% CI		Q_W
				L	*U*	
Flextime	41	2,291	.30	.26	.35	1004.55**
Compressed work week	25	2,921	.29	.23	.34	210.58**

Chapter References

Arthur, W., Bennett, W., & Huffcutt, A. I. (2001). *Conducting meta-analysis using SAS.* Mahwah, NJ: Erlbaum.

Baltes, B. B., Briggs, T. E., Huff, J. W., Wright, J. A., and Neuman, G. A. (1999). Flexible and compressed work-week schedules: A meta-analysis of their effects on work-related criteria. *Journal of Applied Psychology, 84*(4), 496-513.

Glass, G. V. (1976). Primary, secondary and meta-analysis of research. *Educational Researcher, 5,* 3-8.

Hunter, J. E., & Schmidt, F. L. (1990). *Methods of meta-analysis: Correcting error and bias in research findings.* Newbury Park, CA: Sage Publications.

Roth, P. L., BeVier, C. A., Switzer, F. S., & Schippmann, J. S. (1996). Meta-analyzing the relationship between grades and job performance. *Journal of Applied Psychology, 81*(5), 548-556.

Viswesvaran, C., Ones, D. S., & Schmidt, F. L. (1996). Comparative analysis of the reliability of job performance ratings. *Journal of Applied Psychology, 81*(5), 557-574.

Chapter 2
The Meta-Analysis

The meta-analyses presented in this book were conducted using the techniques recommended by Hunter and Schmidt (1990) and Arthur, Bennett, and Huffcutt (2001). Unless specified in later chapters, all meta-analyses were conducted using the following method.

Finding Studies

As mentioned in the introduction, the first step was to locate studies correlating the results of selection methods (e.g., cognitive ability, education, personality) with some measure of law enforcement performance (e.g., academy grades, supervisor ratings). The active search for such studies was concentrated on journal articles, theses, and dissertations published between 1970 and 2003. Studies published prior to 1970 or more recently than 2003 were included when found but inclusion outside of the years 1970-2003 would not be considered exhaustive. To find relevant studies, the following sources were used:

- *Dissertation Abstracts Online* was used to search for relevant dissertations. Interlibrary loan was used to obtain most of the dissertations. When dissertations could not be loaned, they were purchased from the University of Michigan dissertation service. There were a few dissertations and theses that could not be obtained because their home library would not loan them and they were not available for purchase.
- *WorldCat* was used to search for relevant master's theses, dissertations, and books. *WorldCat* is a listing of books contained in many libraries throughout the world and was the single best source for finding relevant master's theses.

- *PsycInfo*, *InfoTrac OneFile*, *ArticleFirst*, *ERIC*, *Periodicals Contents Index*, *Factiva*, *Lexis-Nexis*, and *Criminal Justice Abstracts* were used to search for relevant journal articles and other periodicals.
- Hand searches were made of the *Journal of Police and Criminal Psychology*, *Journal of Criminal Justice*, *Journal of Police Science and Administration*, *Police Quarterly*, and *Public Personnel Management*.
- Reference lists from journal articles, theses, and dissertations were used to identify other relevant material.

Keywords used to search electronic databases included combinations of occupational terms (e.g., police, law enforcement, sheriff) with predictors (e.g., education, personality, MMPI, CPI, cognitive ability, IQ, military), methods (e.g., validity, relationship, predicting), and criteria (e.g., academy, performance, grades, commendations).

The search for documents stopped when computer searches failed to yield new sources and no new sources from reference lists appeared. To be included in this book, a study had to be an empirical investigation of the validity of a selection method applied to a law enforcement sample and had to include data. There were hundreds of articles on the topic of police selection that did not include data and these were not summarized. The literature search yielded 339 relevant studies: 155 journal articles, 41 master's theses, 115 doctoral dissertations, 10 technical reports, 12 books, and 6 conference presentations. The 339 studies included 110,464 law enforcement personnel—88% of whom were men and 72% of whom were white. Further information on the studies used in this book can be found in Table 2.1. Technical summaries of each article included in the meta-analyses can be accessed in a PDF file at www.radford.edu/~maamodt/riles.htm.

To be included in the meta-analysis, an article had to report the results of an empirical investigation and had to include a correlation coefficient, another statistic that could be converted to a correlation coefficient (e.g., t, F, χ^2), or tabular data or raw data that could be analyzed to yield a correlation coefficient. Articles reporting results without the above statistics (e.g., "We found a significant relationship between military experience and academy performance" or "Our experience in hiring military veterans was not positive") could not be included in the meta-analysis. Studies listing personality means but no validity coefficients were included so that a meta-analysis of the "police personality" could be conducted.

Table 2.1
Characteristics of studies used in this project

Characteristic	Number of studies
Source	
Journal article	155
Doctoral dissertation	115
Master's thesis	41
Technical report	10
Book	12
Conference presentation	6
Study Decade	
1950s	2
1960s	17
1970s	65
1980s	107
1990s	120
2000s	28
Law Enforcement Sample	
City police	275
State police	24
Campus police	6
Sheriff's department	12
Military police	3
Security	2
Corrections	5
Traffic officers	4
Transit officers	1
Other	4

Converting Research Findings to Correlations

Once the studies were located, statistical results that needed to be converted into correlation coefficients (r) were done so using the formulas provided in Arthur, Bennett, and Huffcutt (2001). In some cases, raw data or frequency data listed in tables were entered into a SAS or Excel program to directly compute a correlation coefficient. When conversions were made, they were noted in the article summaries located in the previously mentioned PDF file. There were several occasions when the reanalysis of the raw data indicated a computational or transposition error in the original study. For example, the results of one dissertation were clearly different from all other studies. When reviewing the raw data included at the end of the dissertation, it was apparent that the author had coded missing data with a "9" but did not indicate in the SPSS program that a nine represented missing data. When the data were reanalyzed with a nine as missing data rather than as the digit nine, the results were consistent with other studies. When such reanalysis was conducted, a notation was made in the article summary.

Cumulating Validity Coefficients

After the individual correlation coefficients were computed, the validity coefficient for each study was weighted by the size of the sample and the coefficients combined using the method suggested by Hunter and Schmidt (1990) and Arthur, Bennett, and Huffcutt (2001). In addition to the mean validity coefficient, the observed variance, amount of variance expected due to sampling error, 95% confidence interval, true (corrected) correlation, and 90% credibility interval were calculated. All meta-analysis calculations were performed using *Meta-Manager*, an Excel-based program written by the author for this project. The integrity of the formulas in *Meta-Manager* were validated using datasets and meta-analysis results provided in Arthur, Bennett, and Huffcutt (2001) and in Hunter and Schmidt (1990). Copies of the *Meta-Manager* template can be obtained without cost from the author (maamodt@radford.edu).

Correcting for Artifacts

As mentioned in the previous chapter, when conducting a meta-analysis, it is desirable to adjust correlation coefficients to correct for attenuation due to error associated with predictor unreliability, criterion unreliability, restriction of range, and a host of other artifacts (see Hunter & Schmidt, 1990 for a thorough discussion). Corrections for restriction in range and measurement unreliability can be made in one of two ways. The most desirable is to correct the validity coefficient from each study based on the predictor reliability, criterion reliability, and restriction of range associated with that particular study. To do this, however, each study must provide this information and very few of the studies included in this project, or any meta-analysis project for that matter, provide information on range restriction, predictor reliability, or criterion reliability. So, corrections were made to the mean validity coefficients rather than the individual validity coefficients.

The corrections were made using either the mean values from studies providing the necessary information or from values suggested in previous meta-analyses. In conducting a meta-analysis, the figures used to make corrections is an important one because the wrong choice will result in either an overestimation or underestimation of the true validity of a construct. In considering the figures to use in this meta-analysis, I compared the corrections used in previous meta-analyses with those obtained for the studies in this project. Those figures are shown in Tables 2.2, 2.3, and 2.4.

The corrections used in this meta-analysis are shown in Table 2.5. To correct for unreliability of supervisor ratings, I used the reliability data from the 25 studies in this project that provided such information. Though the reliability coefficient of .64 is above the .52 (Viswesvaran, Ones, & Schmidt, 1996) and .51 (Conway & Huffcutt, 1997) reported in previous meta-analyses of the reliability of supervisor ratings, it is lower than the .73 used by Hirsh, Northrop, and Schmidt (1986) in their earlier meta-analysis of the validity of cognitive ability in law enforcement occupations and similar to those used in meta-analyses by Tett, Jackson, and Rothstein (1991), Gaugler, Rosenthal, Thornton, and Bentson (1997), and Salgado, Anderson, Moscoso, Bertua, and De Fruyt (2003a). Though the use of such a high reliability coefficient may result in an underestimation of the true validity of selection techniques in predicting supervisor ratings, it seems reasonable given the consistency of my findings and those of previous meta-analyses.

Table 2.2
Review of Predictor Reliability Estimates

Predictor/Source	K	r_{xx}	SD_{xx}	VAR_{xx}
Cognitive Ability				
Salgado et al. (2003)	31	.83	.0900	.0081
Hirsch et al. (1986)	69	.80		
CURRENT STUDY	56	.82	.1034	.0107
Personality				
Tett et al. (1991)	122	.76	.1030	.0106
Barrick & Mount (1991)		.76	.0800	.0064
Salgado (1997)		.80	.1100	.0121
Barrick et al. (2003)		.89		
CURRENT STUDY		.76	.1100	.0121
Interviews				
McDaniel et al. (1994)	167	.68	.2500	.0625
Conway et al. (1995)		.70	.0350	.0012
Interest Inventories				
Barrick et al. (2003)		.92		
CURRENT STUDY	39	.87	.0484	.0023
Assessment Centers				
Arthur et al. (2003)	37	.86	.0707	.0050

To correct for unreliability of academy grades (training performance), I used the reliability of .84 for grades found by Roth et al. (1996). Because Roth et al. did not provide a standard deviation, the standard deviation of the reliability of training measures from Salgado et al. (2003b) was used. Because academy GPA was the most common training measure in this study, it made sense to use the estimate of grade reliability by Roth et al. rather than the estimate of training reliability by Salgado et al. (2003b) that contained mostly instructor ratings of success. To correct for unreliability in discipline problems and use of force, the coefficient from the extensive meta-analysis on counterproductive behaviors by Ones, Viswesveran, and Schmidt (1993) was used

(r_{yy} = .69, SD_{yy} = .09). To correct for unreliability in activity (e.g., arrests made, citations issued), the productivity coefficients (r_{yy} = .73, SD_{yy} = .14) from Hunter, Schmidt, and Judiesch (1990) were used. Because information was not available from studies in this meta-analysis or in previous meta-analyses, no corrections for criterion unreliability were made for commendations, absenteeism, and injuries and no corrections for predictor unreliability were made for education, military experience, or background problems (e.g., traffic tickets, times arrested).

Table 2.3 Review of Criterion Reliability Estimates

Criterion/Source	Meta-analysis topic	K	r_{yy}	SD_{yy}	VAR_{yy}
Supervisor Ratings of Performance					
Viswesveran et al. (1996)	Rater agreement	40	.52	.0950	.0090
Conway & Huffcutt (1997)	Rater agreement	67	.51	.1480	.0220
Salgado et al. (2003a)	Cognitive ability	19	.52	.1900	.0361
Rothstein (1990)	Rater agreement	79	.48	.0710	.0050
Ones, et al. (1993)	Integrity tests	153	.52	.0900	.0081
Barrick & Mount (1991)	Personality		.52	.0500	.0025
Hurtz & Donovan (2000)	Personality		.53	.1500	.0225
Salgado et al. (1997)	Personality		.62	.1200	.0144
Tett et al. (1991)	Personality		.62	.1500	.0094
Gaugler et al (1987)	Assessment centers		.60	.1265	.0160
Arthur et al. (2003)	Assessment centers	37	.76	.1342	.0180
CURRENT STUDY	Police selection	25	.64	.1600	.0256
Counterproductive Behaviors					
Ones et al. (1993)	Integrity tests	171	.69	.0900	.0081
Activity (Productivity)					
Hunter et al. (1990)			.73	.1400	.0196
Training Success					
Salgado et al. (2003a)	Cognitive ability	15	.56	.0900	.0081
Roth et al. (1996)	Grades		.84		
Hunter & Hunter (1984)	Cognitive ability		.81		
Gaugler et al. (1987)	Assessment centers		.80	.0447	.002

Table 2.4
Range restriction corrections in previous meta-analyses

Meta-analysis	*u*	VAR	SD	K	Meta-analysis Topic
Studies in this project	.46	.0276	.1660	14	Police selection
Hirsh et al. (1986)	.61	.0146	.1210	17	Cognitive ability – police
Salgado et al (2003)	.62	.0625	.2500	20	Cognitive ability
Hunter & Hunter (1984)	.67	.0077	.0877		Cognitive ability
McDaniel et al. (1994)	.68	.0256	.1600	14	Interviews
Huffcutt & Arthur (1994)	.74	.0256	.1600	15	Interviews
Latham & Sue-Chan (1999)	.78	.0361	.1900	3	Situational interview
Ones et al (1993)	.81	.0361	.1900	79	Integrity tests
Gaugler et al. (1987)	.90	.0320	.1789		Assessment centers
Salgado (1997)	.92	.0121	.1100		Personality
Barrick & Mount (1991)	.94	.0025	.0500		Personality
Hurtz & Donovan (2000)	.92	.0729	.2700		Personality
Barrick et al. (2003)	.93				Interest inventories
Median	.78	.0320	.1789		

Determining the appropriate correction for range restriction was difficult. Fourteen studies provided either selection ratios or unrestricted and restricted standard deviations that could be converted into a measure of range restriction (u). As shown in Table 2.4, the value of u used in previous studies has ranged tremendously and all have been substantially higher than the values reported in the studies comprising this meta-analysis. It was not surprising that u was lower in this series of meta-analyses than in previous meta-analyses. In law enforcement selection, there are usually many applicants applying for few openings (low selection ratio) and the multiple selection methods used in each screening (e.g., cognitive ability, interview, background check, personality tests) compounds the number of applicants that would be screened out with only one method as was the case with earlier meta-analyses studying the validity of a single predictor (e.g., cognitive ability). Given the selection ratio and multiple hurdles commonly encountered in law enforcement selection, the low value of u (.46) obtained from the 14 studies providing relevant information seems appropriate. To use a value so different from other meta-analyses, however, would result in unusually high corrected validities. As a compromise, the median *u* (.78) and median standard deviation (.1789) from this and previous meta-analyses were used to correct for range restriction. Such a choice, especially when combined with the conservative correction for unreliability of supervisor ratings, may result in an underestimation of the true validity of the various selection methods used in law enforcement settings. However, I would prefer to be conservative and underestimate the true validity than to overestimate it.

Table 2.5
Corrections for Attenuation Used in this Meta-Analysis

	r	VAR	SD	K	Source
Corrections for Predictor Unreliability					
Cognitive ability	.82	.0107	.1035	57	Studies in this meta analysis
Personality inventories	.76	.0121	.1100		Studies in this meta-analysis
Openness	.73	.0169	.1300	139	Viswesvaran & Ones (2000)
Conscientiousness	.76	.0169	.1300	193	Viswesvaran & Ones (2000)
Extraversion	.78	.0121	.1100	176	Viswesvaran & Ones (2000)
Agreeableness	.75	.0121	.1100	119	Viswesvaran & Ones (2000)
Stability	.73	.0100	.1000	221	Viswesvaran & Ones (2000)
Interest inventories	.87	.0023	.0484	39	Studies in this meta-analysis
Interviews	.68	.0625	.2500	167	McDaniel et al. (1994)
Assessment centers	.86	.0050	.0707	37	Arthur et al. (2003)
Clinical interviews	.73	.0121	.1100	25	McDaniel et al. (1994)
Corrections for Criterion Unreliability					
Academy grades	.84				Roth et al. (1996)
Academy grades		.0081	.0900	15	Salgado et al. (2003)
Supervisor ratings	.64	.0256	.1600	25	Studies in this meta-analysis
Peer ratings	.42	.0112	.1060		Viswesvaran et al. (1996)
Activity (productivity)	.73	.0196	.1400		Hunter et al. (1990)
Discipline problems	.69	.0081	.0900	171	Ones et al. (1993)

Searching for Moderators and Generalizing Results

Being able to generalize meta-analysis findings across all similar organizations and settings (validity generalization) is an important goal of any meta-analysis. In this meta-analysis when variance due to sampling error, range restriction, attenuation in the predictor, and attenuation in the criterion accounted for less than 75% of observed variance, the next step was to remove outliers. Outliers were defined as the validity coefficients that represent the top and bottom 5% of the studies (Huber, 1980; Tukey, 1960). Outliers are removed from meta-analyses because the assumption is that a study obtaining results that are very different from those found in other studies did so due to such factors as calculation errors, coding errors, or the use of a unique sample. After removing outliers, if the variance accounted for was still less than 75%, a search for such potential moderators as the year the study was conducted, agency size (small, medium, large), sample type (police department, sheriff's department, SWAT unit, military police), location (South, Northeast, West, Midwest), percentage of females in the sample, percentage of minorities in the sample, and percentage of officers in the sample with bachelor's degrees was undertaken. The specific moderators investigated for each meta-analysis are listed in subsequent meta-analysis chapters.

Chapter References

Arthur, W., Bennett, W., & Huffcutt, A. I. (2001). *Conducting meta-analysis using SAS.* Mahwah, NJ: Erlbaum.

Arthur, W., Day, E. A., McNelly, T. L., & Edens, P. S. (2003). A meta-analysis of the criterion-related validity of assessment center dimensions. *Personnel Psychology, 56*(1), 125-154.

Barrick, M. R., & Mount, M. K. (1991). The big five personality dimensions and job performance: A meta-analysis. *Personnel Psychology, 44*(1), 1-26.

Barrick, M. R., Mount, M. K., & Gupta, R. (2003). Meta-analysis of the relationship between the five-factor model of personality and Holland's occupational types. *Personnel Psychology, 56*(1), 45-74.

Conway, J. M., & Huffcutt, A. I. (1997). Psychometric properties of multi-source performance ratings: A meta-analysis of subordinate, supervisor, peer, and self-ratings. *Human Performance, 10*(4), 331-360.

Gaugler, B. B., Rosenthal, D. B., Thornton, G. C., & Bentson, C. (1987). Meta-analysis of assessment center validity. *Journal of Applied Psychology, 72*(3), 493-511.

Hirsh, H. R., Northrop, L. C., & Schmidt, F. L. (1986). Validity generalization results for law enforcement occupations. *Personnel Psychology, 39*, 399-420.

Huber, P. J. (1980). *Robust statistics*. New York: Wiley.

Huffcutt, A. I., & Arthur, W. (1994). Hunter and Hunter (1984) revisited: Interview validity for entry-level jobs. *Journal of Applied Psychology, 79*(2), 184-190.

Hunter, J. E., & Hunter, R. F. (1984). Validity and utility of alternative predictors of job performance. *Psychological Bulletin, 96(*1), 72-98.

Hunter, J. E., & Schmidt, F. L. (1990). *Methods of meta-analysis: Correcting error and bias in research findings*. Newbury Park, CA: Sage Publications.

Hunter, J. E., Schmidt, F. L., & Judiesch, M. K. (1990). Individual differences in output variability as a function of job complexity. *Journal of Applied Psychology, 75*(1), 334-349.

Hurtz, G. M., & Donovan, J. J. (2000). Personality and job performance: The big five revisited. *Journal of Applied Psychology, 85*(6), 869-879.

Latham, G.P., & Sue-Chan, C. (1999). A meta-analysis of the situational interview: An enumerative review of reasons for its validity. *Canadian Psychology, 40*, 56-67.

McDaniel, M. A., Whetzel, D. L., Schmidt, F. L., & Maurer, S. D. (1994). The validity of employment interviews: A comprehensive review and meta-analysis. *Journal of Applied Psychology, 79*, 599-616.

Ones, D. S., Viswesvaran, C., & Schmidt, F. L. (1993). Comprehensive meta-analysis of integrity test validities:

Findings for personnel selection and theories of job performance. *Journal of Applied Psychology, 78*(4), 679-703.
Roth, P. L., BeVier, C. A., Switzer, F. S., & Schippmann, J. S. (1996). Meta-analyzing the relationship between grades and job performance. *Journal of Applied Psychology, 81*(5), 548-556.
Rothstein, H. R. (1990). Interrater reliability of job performance ratings: Growth to asymptote level with increasing opportunity to observe. *Journal of Applied Psychology, 7*, 322-327.
Salgado, J. F. (1997). The five-factor model of personality and job performance in the European community. *Journal of Applied Psychology, 82*(1), 30-43.
Salgado, J. F., Anderson, N., Moscoso, S., Bertua, C., & de Fruyt, F. (2003a). International validity generalization of GMA and cognitive abilities: A European Community meta-analysis. *Personnel Psychology, 56*(3), 573-605.
Salgado, J. F., Anderson, N., Moscoso, S., Bertua, C., de Fruyt, F., & Rolland, J. P. (2003b). A meta-analytic study of general mental ability validity for different occupations in the European Community. *Journal of Applied Psychology, 88*(6), 1068-1081.
Tett, R. P., Jackson, D. N., & Rothstein, M. (1991). Personality measures as predictors of job performance: A meta-analytic review. *Personnel Psychology, 44*(4), 703-742.
Tukey, J. W. (1960). A survey of sampling from contaminated distributions. In I. Olkin, J. G. Ghurye, W. Hoeffding, W. G. Madoo, H. Mann (Eds). *Contributions to probability and statistics.* Stanford, CA: Stanford University Press.
Viswesvaran, C., & Ones, D. S. (2000). Measurement error in "Big Five Factors" personality assessment: Reliability generalization across studies and measures. *Educational and Psychological Measurement, 60*, 224-235.
Viswesvaran, C., Ones, D. S., & Schmidt, F. L. (1996). Comparative analysis of the reliability of job performance ratings. *Journal of Applied Psychology, 81*, 557-574.

Chapter 3
Cognitive Ability and Police Performance

In one form or another, cognitive ability tests are commonly used in law enforcement selection. This category of tests includes a wide variety of tests ranging from those tapping general intelligence to those tapping such specific aspects of cognitive ability as reading, math, vocabulary, and logic. From a content validity perspective, cognitive ability tests are thought to be important in law enforcement selection as they are related to the ability to perform such tasks as learning and understanding new information, writing reports, making mathematical calculations during investigations, and solving problems.

The purpose of this chapter is to first describe the cognitive ability tests commonly used in studies on law enforcement selection and then report the results of the meta-analysis investigating the relationship between cognitive ability and academy, probationary, and on-the-job performance. This meta-analysis will serve to update a similar meta-analysis published by Hirsh, Northrop, and Schmidt (1986) and complement a meta-analysis of the validity of cognitive ability tests in Europe published by Salgado, Anderson, Moscoso, Bertua, de Fruyt, and Rolland (2003).

Cognitive Ability Tests Used in Law Enforcement Validation Research

Cognitive ability tests can be placed into four categories on the basis of where they were developed and the extent to which they are commercially available.

Publisher Developed General Cognitive Ability Tests

This first category includes cognitive ability tests developed by national test publishers. These tests are available for a wide variety of uses and can be purchased directly from the publishers. Though these tests were not specifically designed for law enforcement selection, many are certainly compatible with constructs related to law enforcement performance. Tests from this category commonly used in law enforcement selection include the following:

Nelson-Denny Reading Test

The Nelson-Denny Reading Test is a 118-item test of reading comprehension and vocabulary that takes 45 minutes to complete. According to the test manual, the test-retest reliability is .77, internal reliability is .96, and alternate form reliability is .90.

Research by Rose (1995) indicated that a high-school reading level was needed in the academy and that Nelson-Denny scores significantly correlated (r = .59) with academy grades. Whitton (1990) found a correlation of .48 with academy grades. Greb (1982) reported a median correlation of .52 and Spaulding (1980) reported correlations of .55, .38, and .37 between Nelson-Denny scores and academy grades. The validity of the Nelson-Denny in predicting patrol ratings was investigated by Spielberger et al. (1979) who found a correlation of .22 for males and .31 for females between Nelson-Denny scores and supervisor ratings of performance. Surrette, Aamodt, and Serafino (1990) did not find a significant relationship (r = .05) between performance ratings and Nelson-Denny scores.

Otis-Lennon School Ability Test

The Otis-Lennon was designed to tap abstract thinking and reasoning ability in children. For law enforcement selection, Level G (grades 9-12) should be used. Level G takes 60 minutes to complete and can be administered either individually or in a group setting.

Reviews of the Otis-Lennon indicate that the internal reliability is about .90 for the total score and a bit lower for the verbal and nonverbal scores. Research using the Otis-Lennon has found correlations of .70 (Kleiman & Gordon, 1986) and .33 (Gordon & Kleiman, 1976) with academy grades, .34 (McAlister, 1970) with academy graduation, and .14 (Kleiman & Gordon, 1986) and .18 (Talley & Hinz, 1990) with supervisor ratings of job performance.

Watson-Glaser Critical Thinking Appraisal

The Watson-Glaser taps five areas of critical thinking: inferences, deductions, recognition of assumptions, interpretation, and evaluation of arguments. Two formats of the test are available; one with 80 items and the other with 100 items. The test is untimed, can be administered in a group setting, and takes about 50 minutes to complete.

According to the test manual, the Watson-Glaser has a test-retest reliability of .73, an alternate-form reliability of .75, and an internal reliability of .77. The test correlates .41 with the WAIS, .60 with the Otis Mental Ability Test, and .54 with the verbal portion of the Scholastic Aptitude Test (SAT).

Champion (1994) reported a significant validity coefficient of .52 in predicting police academy performance, and a nonsignificant coefficient of .01 in predicting police patrol performance. The mean score for police officers was 55.4 in the one study listed in the test manual and 52 in the study by Champion (1994). These means are lower than the mean of 59.5 for college seniors and higher than the mean of 46.6 for high school seniors.

Wechsler Adult Intelligence Scale (WAIS)

The WAIS is the most commonly used cognitive ability test in clinical psychology. The WAIS is administered individually by a qualified psychologist. This individual administration takes about an hour and a half at a cost of $100-$200 per applicant. Though the WAIS is an outstanding test for many clinical uses, due to its lengthy administration time and it's cost, it is not an ideal test for law enforcement selection. This is especially true given that there are other less expensive tests available that are equally valid in predicting law enforcement performance. In the only two studies found using the WAIS, Brewster and Stoloff (2003) found a correlation of .38 between WAIS scores and performance ratings of 71 police officers after one year on the job and Kenny and Watson (1990) found a correlation of .61 with academy grades.

Wonderlic Personnel Test

The Wonderlic Personnel Test is probably the most commonly used cognitive ability test in industry. Its popularity is in no small part due to the fact that it is a timed test that only takes 12 minutes to complete, is relatively inexpensive (about $3.00 per applicant), can be quickly hand scored, is extensively normed (including norms for law enforcement), and correlates highly with such well known tests as the WAIS (r = .92). The Wonderlic consists of 50 items for which applicants write their answers on the side of the page. These items tap vocabulary, analogies, math, and logic.

Research has shown that Wonderlic scores have a test-retest reliability of .88, an internal consistency (KR-20) reliability of .88, and an alternate form reliability of .84. Correlations of .19 (Super, 1995) and .06 (Hankey, 1968) with patrol performance and .28 with academy performance (Hankey, 1968) have been reported. The average score of 1,854 police applicants taking the Wonderlic is 20.93 with a standard deviation of 6.14.

Shipley Institute of Living Scale

The Shipley is a 60-item test measuring vocabulary and abstract thinking. The test can be group administered, is timed, and takes 20 minutes to complete. The Shipley was developed to measure intellectual ability and impairment in individuals 14 years and older. As such, it is probably more useful in clinical work than it is as a method of selection for law enforcement personnel.

Reliabilities for the vocabulary, abstraction, and total score are .87, .89, and .92 respectively. The Shipley correlates about .70 with the Wechsler-Bellevue Scale (two studies found correlations of .77 and .65).

In the studies investigating the validity of the Shipley in a law enforcement context, Surrette, Aamodt, and Serafino (1990) found no correlation between Shipley scores and probationary performance (r = .00), however Gardner (1994) found a correlation of .16 and Scogin, Schumacher, Gardner, and Chaplin (1995) a correlation of .39 between Shipley scores and performance ratings, Davis and Rostow (2003) found a significant negative correlation (r = -.09) with being fired for cause, and Mullins and McMains (1996) found a significant correlation with academy performance (r = .50).

Nationally Developed Law Enforcement Tests

The second category of cognitive ability tests includes those developed by consultants or trade organizations (e.g., IPMA, IACP) for specific use with law enforcement agencies.

IPMA

The IPMA is sold by the International Public Management Association and is commonly known as the A-3. This test has 100 items tapping verbal ability as well as the ability to learn and remember details, follow instructions, and use judgment and knowledge. The test can be administered in a group setting, has a 130-minute time limit, and costs a little over $10.00 per applicant. The A-3 correlates .20 with supervisor ratings and has an internal reliability of .83 (IPMA, 1992). Women score slightly higher than men (d = .10) and whites score higher than African Americans (d = .47).

Police Officer Selection Test (POST)

The POST is published by Stanard and Associates, a consulting firm located in Chicago. The POST consists of 75 questions tapping math, reading, grammar, and writing skills. The test is timed and takes about 90 minutes to complete including the 75 minutes for the test and 15 minutes for instructions. Rafilson and Sison (1996) report that the POST has an internal reliability of .79 and a validity of .54 in predicting academy performance.

Law Enforcement Candidate Record (LECR)

The Law Enforcement Candidate Record (LECR) is published by Richardson, Bellows, & Henry (1989) and is a combination of a cognitive ability test and a biodata instrument. The test-retest reliability for LECR total scores has been reported at .79 (Erwin & Mead, 1997). In two extensive studies reported in the technical manual and supporting materials, the cognitive component of the LECR correlated .16 with supervisor ratings of performance and the combination of cognitive ability and biodata correlated .28 and .33 with supervisor ratings of performance.

Law Enforcement Selection Inventory

The Law Enforcement Selection Inventory is published by Personnel Research Associates, a consulting firm located in Pulaski, Virginia. The test consists of 80 items tapping math, grammar, vocabulary, and logic. Because each of the items uses a law enforcement example or context, the test is highly face-valid. The test has a test-retest reliability of .87 and correlates .70 with the WAIS, .72 with the Nelson-Denny, and .72 with the IPMA A-3. Research on the Law Enforcement Selection Inventory has shown that it correlated .43 with grades in the academy.

Tests Developed by the Federal Government

The third category of tests was developed by the Federal government for use either with the military or with general employment testing. The major test in this category used in previous law enforcement research is the Army General Classification Test.

Army General Classification Test (AGCT)

The AGCT was developed by the Army in 1940 to aid in the selection and placement of military recruits. A civilian version of the AGCT was released in 1960 and was used by many law enforcement agencies in the 1960's and early 1970's. Though it is no longer commonly used in the civilian sector, it will be discussed here because it was the cognitive ability measure used in several validity studies.

The AGCT contains 150 items tapping skills in vocabulary, math, and spatial relations. The test has an internal reliability of .97, a test-retest reliability of .82, and an alternate forms reliability of .90. The AGCT correlates .83 with the Otis and .83 with the WAIS.

Research using the AGCT in a law enforcement context has been positive. Correlations between AGCT scores and academy performance of .38 (Hess, 1972), .52 (Dubois & Watson, 1950), .58 (Mullineaux, 1965), .69 (Clopton, 1971), .16 (Clopton, 1971), and .60 (Mills, McDevitt, & Tonkin, 1966) have been reported in the literature. Correlations with actual patrol performance have not been positive as Hess (1972) found a correlation of -.12, Dubois and Watson (1950) a correlation of .10, and Clopton (1971) correlations of .14 and -.46 in two studies with small samples.

Locally Developed Civil Service Exams

The fourth category consists of cognitive ability tests developed by various municipalities and Civil Service Commissions for their own use. Though it is not unusual for these tests to be shared with other agencies, they are not commercially available.

In validation research with locally developed tests, correlations of .35 (Abbatiello (1969), .38 (Boehm, Honey, & Kohls,1983), .62 (Cortina, Doherty, Schmitt, Kaufman, & Smith., 1995), .43 (Giannoni, 1979), .24 (Gruber,1986), .27 (Hausknecht, Trevor, & Farr, 2002), .60 (Palmatier, 1996), .20 (Plummer, 1979), .51 (Scarfo, 2002), .18 (Spurlin & Swander, 1987), .36 (Staff,

1992), and .42 (Wexler & Sullivan, 1982) were found between civil service exams and academy grades. Correlations of .30 (Bertram, 1975), .30 (Cortina et al., 1995), .14 (Dayon et al., 2002), .17 (Giannoni, 1979), .19 (Gruber, 1986), .16 (Palmatier, 1996), .37 (Spurlin & Swander, 1987), and .17 (Wexler & Sullivan, 1982) have been reported between civil service exam scores and performance ratings.

Results and Discussion

As shown in Table 3.1, cognitive ability was significantly related to grades in the police academy (r = .41, ρ = .62), supervisor ratings of patrol performance (r = .16, ρ = .27), activity (r = .19, ρ = .33), and serious discipline problems (r = - .12, ρ = - .21). Validity coefficients for minor discipline problems such as complaints and reprimands (r = .03, ρ = .06), injuries (r = - .06, ρ = - .08), absenteeism (r = -.03, ρ = - .05), and commendations (r = -.01, ρ = - .02) were not considered statistically significant as their confidence intervals include zero. As shown in Tables 3.2 and 3.3, reading tests and tests designed especially for law enforcement selection purposes seem to be slightly better predictors than the other types of cognitive ability tests.

The correlations between cognitive ability and academy performance (r = .41) and patrol performance (r = .16) are a bit higher, but similar to, the correlations found for academy performance (r = .34) and patrol performance (r = .09) in a previous meta-analysis on cognitive ability and law enforcement performance in the United States by Hirsch, Northrup, and Schmidt (1986). The correlation with performance ratings was similar to the correlation found in a meta-analysis by Salgado et al. (2003b) of only five studies in the European Community (r = .12). The correlation with training performance found in this meta-analysis was much higher than that found across three European Community studies by Salgado et al. (r = .13) Interestingly, the mean validity coefficients from this meta-analysis and the one by Hirsch and her colleagues are lower than those reported by Hunter and Hunter (1984) in their meta-analysis of the validity of cognitive ability across all jobs.

After correcting for attenuation in the predictor and criterion as well as range restriction, two observations stand out. First, almost all of the variability in validity coefficients across studies can be attributed to sampling error and measurement artifacts. For the two criteria in which study artifacts did not explain at east 75% of the variance (injuries, less severe disciplinary problems), the number of studies was too small to search for moderators. Second, the corrected validity coefficients of .62 for academy grades and .27 for supervisor ratings of patrol performance are of impressive magnitude.

The results of this meta-analysis indicate that cognitive ability tests are valid predictors of academy and on-the-job performance of law enforcement personnel. Though this meta-analysis demonstrated the validity of cognitive ability in the selection of law enforcement personnel, several questions remain that can only be answered through additional studies. Perhaps the most important of these questions involves the type of cognitive ability being tapped. Most of the studies in this analysis utilized tests tapping several dimensions of cognitive ability (e.g., math, vocabulary, logic) that were combined into a total score whereas other studies used tests of a single construct (e.g., reading, critical thinking). Further research is necessary to compare the validity of general cognitive ability tests with the validity of their more specific counterparts.

Table 3.1
Meta-analysis results for cognitive ability

Criterion	K	N	r	95% Confidence Interval		ρ	90% Credibility Interval		Var	Q_w
				Lower	Upper		Lower	Upper		
Academy Grades	61	14.437	.41	.33	.48	.62	.47	.78	78%	77.82
Supervisor Ratings	61	16,231	.16	.12	.20	.27	.16	.38	80%	76.40
Commendations	7	2,015	-.01	-.06	.03	-.02	-.07	.04	91%	7.71
Activity	6	656	.19	.11	.27	.33	.33	.33	100%	5.56
Absenteeism	5	1,402	-.03	-.08	.02	-.05	-.05	-.05	100%	2.11
Injuries	3	1,891	-.06	-.13	.02	-.08	-.28	.16	18%	16.35
Discipline Problems	13	4,850	-.06	-.12	.00	-.11	-.36	.18	26%	49.94*
Fired or suspended	7	3,019	-.12	-.15	-.08	-.21	-.21	-.21	100%	6.74
Complaints/reprimands	6	1,831	.03	-.04	.10	.06	-.21	.30	31%	19.54*

K=number of studies, N=sample size, r = mean correlation, ρ = mean correlation corrected for range restriction, criterion unreliability, and predictor reliability, VAR = percentage of variance explained by sampling error and study artifacts, Q_w = the within group heterogeneity

Table 3.2
Validity for academy grades by type of cognitive ability test

Type of Test	K	N	r	95% Confidence Interval		ρ	90% Credibility Interval		Var	Q_w
				Lower	Upper		Lower	Upper		
Academy Grades	62	14,474	.41	.33	.48	.62	.47	.78	78%	77.82
Police tests	9	848	.54	.47	.61	.79	.79	.79	100%	5.04
Reading tests	10	3,340	.50	.45	.55	.75	.75	.75	100%	3.60
General IQ	27	6,899	.45	.38	.52	.67	.67	.67	100%	25.73
Writing tests	5	712	.40	.31	.48	.61	.61	.61	100%	4.28
Federal tests	9	1,810	.37	.29	.44	.57	.44	.70	83%	10.89
General IQ	18	3,359	.12	.07	.17	.21	.09	.32	80%	22.48
Federal tests	6	835	.05	- .04	.13	.08	- .14	.31	54%	11.09
Civil service tests	10	2,736	.19	.14	.24	.33	.33	.33	100%	9.01

K=number of studies, N=sample size, r = mean correlation, ρ = mean correlation corrected for range restriction, criterion unreliability, and predictor reliability, VAR = percentage of variance explained by sampling error and study artifacts, Q_w = the within group heterogeneity

Table 3.3
Validity for supervisor ratings of performance by type of cognitive ability test

Type of Test	K	N	r	95% Confidence Interval		ρ	95% Credibility Interval		Var	Q_w
				Lower	Upper		Lower	Upper		
Supervisor Ratings	61	16,231	.16	.12	.20	.27	.16	.38	80%	76.40
Police tests	9	5,547	.18	.15	.21	.31	.31	.31	100%	5.29
Reading tests	8	1,399	.16	.11	.21	.28	.28	.28	100%	5.91
General IQ	18	3,359	.12	.07	.17	.21	.09	.32	80%	22.48
Writing tests	3	358	.11	.00	.21	.18	.18	.18	100%	0.41
Federal tests	6	835	.05	- .04	.13	.08	- .14	.31	54%	11.09*
General IQ	10	2,736	.19	.14	.24	.33	.33	.33	100%	9.01

K=number of studies, N=sample size, r = mean correlation, ρ = mean correlation corrected for range restriction, criterion unreliability, and predictor reliability, VAR = percentage of variance explained by sampling error and study artifacts, Q_w = the within group heterogeneity

Chapter References

*Aamodt, M. G. (1997). *Technical Manual for the Law Enforcement Selection Inventory.* Pulaski, VA: Personnel Research Associates, Inc.

*Abbatiello, A. A. (1969). *A study of police candidate selection.* Paper presented at the 77th Annual Convention of the American Psychological Association, Washington, D.C.

*Barbas, C. (1992). *A study to predict the performance of cadets in a police academy using a modified cloze reading test, a civil service aptitude test, and educational level.* Unpublished doctoral dissertation, Boston University.

*Bertram, F. D. (1975). *The prediction of police academy performance and on-thee-job performance from police recruit screening measures.* Unpublished doctoral dissertation, Marquette University.

*Black, J. (2000). Personality testing and police selection: Utility of the "Big Five." *New Zealand Journal of Psychology, 29*(1), 2-9.

*Boehm, N. C., Honey, R., & Kohls, J. (1983). Predicting success in academy training: The POST reading and writing test battery. *Police Chief, 50*(10), 28-31.

*Brewster, J., & Stoloff, M. (2003). Relationship between IQ and first-year performance as a police officer. *Applied H.R.M. Research, 8*(1), 49-50.

*Campa, E. E. (1993). *The relationship of reading comprehension and educational achievement levels to academy and field training performance of police cadets.* Unpublished doctoral dissertation, Texas A&M University.

*Cascio, W. F. (1977). Formal education and police officer performance. *Journal of Police Science and Administration, 5*(1), 89-96.

*Champion, D. H. (1994). *A study of the relationship between critical thinking levels and job performance of police officers in a medium size police department in North Carolina.* Unpublished doctoral dissertation, North Carolina State University.

*Clopton, W. (1971). *Comparison of ratings and field performance data in validating predictions of patrolman performance: A five-year follow-up study.* Unpublished master's thesis, University of Cincinnati.

*Cortina, J. M., Doherty, M. L., Schmitt, N., Kaufman, G., & Smith, R. G. (1995). *Validation of the IPI and MMPI as predictors of police performance.* Paper presented at the annual meeting of the Society for Industrial Organizational Psychology.

*Daley, R. E. (1978). *The relationship of personality variables to suitability for police work.* Unpublished doctoral dissertation, Florida Institute of Technology.

*Davidson, N. B. (1975). *The predictive validity of a police officer selection program.* Unpublished master's thesis, Portland State University.

*Davis, R., & Rostow, C. (2003). Relationship between cognitive ability and background variables and disciplinary problems in law enforcement. *Applied H.R.M. Research, 8*(2), 77-80.

*Dayan, K., Kasten, R., & Fox, S. (2002). Entry-level police candidate assessment center: An efficient tool or a hammer to kill a fly? *Personnel Psychology, 55*(4), 827-849.

*Dibb, G. S. (1978). *A cross-validated comparison of models for the prediction of academy performance and job tenure of police officer recruits.* Unpublished doctoral dissertation, University of Hawaii.

*DuBois, P. H., & Watson, R. I. (1950). A longitudinal predictive study of success and performance of law enforcement officers. *Journal of Applied Psychology, 34*, 90-95.

*Ellison, K. W. (1986). Development of a comprehensive selection procedure for a medium-sized police department. In Reese, J. T. & Goldstein, H. A. (Eds). *Psychological services for law enforcement*, pp 23-27. Washington, D.C.: U.S. Government Printing Office.

Erwin, F. W., & Mead, A. D. (1997). The law enforcement candidate record (LECR). *Security Journal, 8*, 113-116.

*Feehan, R. L. (1977). *An investigation of police performance utilizing mental ability selection scores, police academy training scores, and supervisory ratings of the job performance of police officers.* Unpublished doctoral dissertation, Georgia Institute of Technology.

*Flynn, J. T., & Peterson, M. (1972). The use of regression analysis in police patrolman selection. *Journal of Criminal Law, 63*(4), 564-569.

*Ford, J. K., & Kraiger, K. (1993). Police officer selection validation project: The multijurisdictional police officer examination. *Journal of Business and Psychology, 7*(4), 421-429.

*Garber, C. R. (1983). *Correlation studies using entry scores, training test results, and subsequent job performance ratings of students of the security police academy, Lackland AFB, Texas.* Unpublished doctoral dissertation, Brigham Young University.

*Gardner, J. F. (1994). *The predictive validity of psychological testing in law enforcement.* Unpublished master's thesis, University of Alabama.

*Giannoni, R. J. (1979). *Personnel selection procedures and their relationship with academy training and field performance of state traffic officers.* Unpublished master's thesis, California State University, Sacramento.

*Gordon, M. E., & Kleiman, L. S. (1976). The prediction of trainability using a work sample test and an aptitude test: A direct comparison. *Personnel Psychology, 29*, 243-253.

*Gottlieb, M. C., & Baker, C. F. (1974). Predicting police officer effectiveness. *The Journal of Forensic Psychology, 6*, 35-46.

*Greb, J. T. (1982). *The Nelson-Denny Reading Test as a predictor of police recruit training success.* Unpublished doctoral dissertation, Florida Atlantic University.

*Gruber, G. (1986). The police applicant test: A predictive validity study. *Journal of Police Science and Administration, 14*(2), 121-129.

*Hamack, R. F. (1988). *Pre-academy placement in the Washington State Patrol: Factors associated with* *academy and job performance*. Unpublished master's thesis, Central Washington University.

*Hankey, R. O. (1968). *Personality correlates in a role of authority: The police*. Unpublished doctoral dissertation, University of Southern California.

*Hausknecht, J. P., Trevor, C. O., & Farr, J. L. (2002). Retaking ability tests in a selection setting: Implications for practice effects, training performance, and turnover. *Journal of Applied Psychology, 87*(2), 243-254.

*Henderson, N. D. (1979). Criterion-related validity of personality and aptitude scales. In Spielberger, C. D. (Ed.) *Police selection and evaluation*. New York: Praeger.

*Hess, L. R. (1972). *Police entry tests and their predictability of score in police academy and subsequent job performance*. Unpublished doctoral dissertation, Marquette University.

Hirsh, H. R., Northrop, L. C., & Schmidt, F. L. (1986). Validity generalization results for law enforcement occupations. *Personnel Psychology, 39*(2), 399-420.

*Hooper, M. K. (1988). *Relationship of college education to police officer job performance*. Unpublished doctoral dissertation, Claremont Graduate School.

Hunter, J. E., & Hunter, R. F. (1984). Validity and utility of alternative predictors of job performance. *Psychological Bulletin, 96(*1), 72-98.

*International Personnel Management Association (1992). *Development and validation of the A-3 police officer examination*. Alexandria, VA: author.

*Jayraj, E. A. S. (1984). *A predictive validity study of police officer selection*. Unpublished master's thesis, Southern Connecticut University.

*Jeanneret, P. R., Moore, J. R., Blakley, B. R., Koelzer, S. L., & Menkes, O. (1991). *Development and validation of trooper physical ability and cognitive ability tests: Final report submitted to the Texas Department of Public Safety*. Houston, TX: Jeanneret & Associates.

*Kenney, D. J., & Watson, S. (1990). Intelligence and the selection of police recruits. *American Journal of Police, 9*(4), 39-64.

*Kleiman, L. S. (1978). *Ability and personality factors moderating the relationships of police academy training performance with measures of selection and job performance.* Unpublished doctoral dissertation, University of Tennessee, Knoxville.

*Kleiman, L. S., & Gordon, M. E. (1986). An examination of the relationship between police training academy performance and job performance. *Journal of Police Science and Administration, 14*(4), 293-299.

*Knights, R. M. (1976). *The relationship between the selection process and on-the-job performance of Albuquerque police officers.* Unpublished doctoral dissertation, University of New Mexico.

*McAllister, J. A. (1970). A study of the prediction and measurement of police performance. *Police*, March-April, 258-264.

*McEuen, O. L. (1981). *Assessment of some personality traits tht show a relationship to academy grades, being dismissed from the department, and work evaluation ratings for police officers in Atlanta, Georgia.* Unpublished doctoral dissertation, The Fielding Institute.

*Mealia, R. M. (1990). *Background factors and police performance.* Unpublished doctoral dissertation.

*Mills, A. (1990). *Predicting police performance for differing gender and ethnic groups: A longitudinal study.* Unpublished doctoral dissertation, California School of Professional Psychology.

*Mills, R. B., McDevitt, R. J., & Tonkin, S. (1966). Situational tests in metropolitan police recruit selection. *Journal of Criminal Law, Criminology, and Police Science, 57*(1), 99-106.

*Mullineaux, J. E. (1965). An evaluation of the predictors used to select patrolmen. *Public Personnel Review, 16*, 84-86.

*Mullins, W. C., & McMains, M. (1996). Predicting patrol officer performance from a psychological test battery: A predictive validity study. *Journal of Police and Criminal Psychology*, *10*(4), 15-25.

*Palmatier, J. J. (1996). *The big-five factors and hostility in the MMPI and IPI: Predictors of Michigan State Trooper job performance.* Unpublished doctoral dissertation, Michigan State University.

*Plummer, K. O. (1979). *Pre-employment factors that determine success in the police academy.* Unpublished doctoral dissertation, Claremont Graduate College.

*Pynes, J. (1988). *The predictive validity of an assessment center for the selection of entry-level law enforcement officers.* Unpublished doctoral dissertation, Florida Atlantic University.

*Rafilson, F., & Sison, R. (1996). Seven criterion-related validity studies conducted with the national Police Officer Selection Test. *Psychological Reports*, *78*, 163-176.

*Richardson, Bellows, & Henry (1989). *The RBH Law Enforcement Candidate Record technical report.* Washington, D.C.: Author

*Ronan, W. W., Talbert, T. L., & Mullett, G. M. (1977). Prediction of job performance dimensions: Police officers. *Public Personnel Management*, *6*(3), 173-180.

*Rose, J. E. (1995). *Consolidation of law enforcement basic training academies: An evaluation of pilot projects.* Unpublished doctoral dissertation, Northern Arizona University.

Salgado, J. F., Anderson, N., Moscoso, S., Bertua, C., de Fruyt, F., & Rolland, J. P. (2003b). A meta-analytic study of general mental ability validity for different occupations in the European Community. *Journal of Applied Psychology, 88*(6), 1068-1081.

*Scarfo, S. (2002). Relationship between police academy performance and level of education. *Applied H.R.M. Research, 7*(1), 24.

*Schaller, G. R. (1990). *Use of the Armed Services Vocational Aptitude Battery to predict training outcomes in female military police trainees.* Unpublished doctoral dissertation, Auburn University.

*Schroeder, D. J. (1973). *A study of the validity of the entrance examination for the position of patrolman under the guidelines established by the Equal Opportunity Employment Commission.* Unpublished master's thesis, John Jay College.

*Scogin, F., Schumacher, J., Gardner, J., & Chaplin, W. (1995). Predictive validity of psychological testing in law enforcement settings. *Professional Psychology: Research and Practice, 26*(1), 68-71.

*Shaver, D. P. (1980). *A descriptive study of police officers in selected towns of Northwest Arkansas.* Unpublished doctoral dissertation, University of Arkansas.

*Spaulding, H. C. (1980). *Predicting police officer performance: The development of screening and selection procedures based on criterion-related validity.* Unpublished master's thesis, University of South Florida.

*Spielberger, C. D., Spaulding, H. C., Jolley, M. T., & Ward, J. C. (1979). Selection of effective law enforcement officers: The Florida police standards research project. In Spielberger, C. D. (Ed.) *Police selection and evaluation.* New York: Praeger.

*Spurlin, O., & Swander, C. (1987). *Validity and fairness of the police officer written exam.* Research finding. Seattle, WA: Public Safety Civil Service Commission.

*Staff, T. G. (1992). *The utility of biographical data in predicting job performance: Implications for the selection of police officers.* Unpublished doctoral dissertation, University of Toledo.

*Super, J. T. (1995). Psychological characteristics of successful SWAT/tactical response team personnel. *Journal of Police and Criminal Psychology, 11*(1), 60-63.

*Surrette, M. A., Aamodt, M. G., & Serafino, G. (1990). *Validity of the New Mexico police selection battery.* Paper presented at the annual meeting of the Society of Police and Criminal Psychology, Albuquerque, New Mexico.

*Talley, J. E., & Hinz, L. D. (1990). *Performance prediction of publis safety and law enforcement personnel.* Springfield, IL: Charles C. Thomas.

*Tomini, B. A. (1995). *The person-job fit: implications of selecting police personnel on the basis of job dimensions, aptitudes, and personality traits*. Unpublished doctoral dissertation, University of Windsor.

*Tompkins, L. P. (1977). *An evaluation of police academy training upon selected recruits and its relationship to job performance*. Unpublished master's thesis, Rollins College, FL.

*Truxillo, D. M., Bennett, S. R., & Collins, M. L. (1998). College education and police job performance: A ten-year study. *Public Personnel Management, 27*(2), 269-280.

*Tyler, T. A. (1989). *Executive summary: MEAS police officer examination*. Flossmoor, IL: Merit Employment Assessment Services, Inc.

Viswesvaran, C., Ones, D. S., & Schmidt, F. L. (1996). Comparative analysis of the reliability of job performance ratings. *Journal of Applied Psychology, 81*(5), 557-574.

*Waugh, L. (1996). *Police officer recruit selection: Predictors of academy performance*. Queensland, Australia: Queensland Police Academy.

*Wexler, N., & Sullivan, S. M. (1982). *Concurrent validation of a prototype selection test for entry-level police officer.* Trenton, NJ: New Jersey Department of Civil Service, Division of Examinations.

*Whitton, W. M. (1990). *The Nelson-Denny Reading Test as a predictor of academic performance of police recruits and the impact of nine related variables on recruit academic performance*. Unpublished doctoral dissertation, The Union Institute.

* indicates that the reference was used in the meta-analysis

Chapter 4
Education and Police Performance

The relationship between education and police performance has long been studied by law enforcement experts. As early as 1967, the law enforcement and criminal justice community examined the need to strengthen the academic requirements to become a police officer. One of their primary arguments was that as the general population becomes more intelligent, police officers should maintain or exceed the level of intelligence of the average citizen they serve. Another argument was that to move toward the "professionalization" of the police department, a four-year college degree is a necessary requirement (Trojanowicz & Nicholson, 1976).

In the past four decades, three primary law enforcement commissions have suggested that a four-year college degree be a minimum requirement for becoming a police officer. The first of these commissions, the President's Commission on Law Enforcement and Administration (COLEA), met in 1967. COLEA gave three reasons for this recommendation: performance of police personnel will not improve until educational standards are raised; advanced education assists an officer's ability to make difficult judgments in unpredictable situations; and higher education promotes knowledge of society and human behavior.

The second commission, the National Advisory Commission on Criminal Justice Standards and Goals, in 1973 agreed with COLEA that all officers should have four-year degrees by 1982. These standards were prompted by two factors. First, a high school diploma was no longer a benchmark of superior education, and second, the police officer must remain as intelligent as the general population whose educational standards are rising. The professionalization of police departments involves recognizing the complex nature of policing duties. Therefore, advanced education can only improve performance on these tasks and can help establish greater legitimacy of law enforcement as a profession.

In 1988, the Police Executive Research Forum (PERF) became the third commission to state that police departments should require a four-year degree as a minimum requirement. PERF based their findings on research that found that only 22.6% of all law enforcement personnel had a four-year college degree and over one-third (34.8%) had not attended college (Carter & Sapp, 1992). The study also showed that only 14% of all police departments required a college education for employment. Despite these findings, the PERF study could not produce conclusive evidence that a college education results in better police performance.

Over the last two decades there have been many attempts to justify educational requirements. These attempts fall under four main categories: support from case law, support from qualitative research, support from studies focusing on characteristics of educated police officers, and support from empirical research directly investigating the relationship between education and performance.

Support from Case Law

Several court cases have supported education requirements for law enforcement positions (Scott, 1986). In *League of United Latin American Citizens v. Santa Ana* (1976), *United States v. Buffalo*, *Morrow v. Diolard*, and *Castro v. Beecher* (1972) the courts supported the validity of the minimum requirement of a high school diploma. In *Davis v. Dallas* (1985), the U.S. Court of Appeals for the Fifth Circuit upheld the Dallas Police Department's requirement that applicants have at least 45 hours of college credit. The U.S. Court of Appeals for the Tenth Circuit upheld a department's requirement of college credits and ultimately a bachelor's degree to be eligible for promotion (*Chicano Police Officer's Association v. Stover*, 1976). In *Ice v. Arlington County* (1977), a U.S. District Court ruled that providing pay incentives for officers with a college education was legal and even stated that "All of the evidence disclosed that college trained policemen make better policemen."

Support from Qualitative Research

There have been many qualitative articles written both supporting and refuting the importance of education. For example:

- Buracker (1977) discussed the advantages and disadvantages of educational requirements and called for further research before drawing any conclusions.
- Gross (1973), Miller and Fry (1978), and Tucker and Hyder (1978) questioned the need for a college degree.
- Geary (1970) supported the importance of college education by reporting that when Ventura, California adopted a bachelor's degree requirement, both the crime rate and police turnover decreased.
- Walker (1994) recommended that large agencies require an associate's degree at hire and a bachelor's degree within five years of being on the force.

Support from Research on Characteristics of Educated Officers

Some researchers have justified the use of education requirements by demonstrating that educated officers possess certain desirable characteristics that less educated officers do not. For example, compared to their less educated counterparts, college educated officers:

- are less authoritarian and dogmatic (Dorsey, 1994; Genz & Lester, 1977; Goldstein, 1977; Feldman & Newcomb, 1969; Roberg, 1978; Smith, Locke, & Fenster, 1970)
- use better discretion (Finckenauer, 1975)
- communicate better (Scott, 1986)
- write better reports (Michals & Higgins, 1994; Smith & Aamodt, 1996)
- are more positive about community policing issues (Ferrell, 1994)

Support from Empirical Research

There have been many studies, usually doctoral dissertations, that have empirically investigated the relationship between education and police performance. Unfortunately, many of these articles report findings that cannot be included in a meta-analysis because they did not report their statistical results or did not use a statistic that can be converted into an effect size. For example:

- In a study of 210 officers in the Baltimore Police Department, Finnigan (1976) found a significant relationship between education and performance. He also found that criminal justice majors performed equally to other majors and that social science majors outperformed business majors.
- A study of 418 Michigan State Police Troopers found that troopers with an associate's degree or higher performed better in the academy than troopers with high school degrees, troopers with a bachelor's degree had higher job performance ratings than troopers with high school degrees, and troopers with a criminal justice degree performed equally to other majors (Weirman, 1978).
- Daniel (1982) studied 10 police departments in St. Louis County, Missouri and found that officers with high school diplomas had more than twice as many absences as their counterparts with college degrees.
- Worden (1990) found that education was not related to officer performance in police-citizen encounters.
- Reming (1988) found no education differences between supercops and average cops.
- Smith and Ostrom (1974) reported no positive relationship between education and police attitudes.

This chapter reports the results of a meta-analysis investigating the relationship between education and various aspects of police performance. This chapter will address four questions:

1) Is education a valid predictor of police performance?

2) Is education as valid a predictor early in an officer's career as later?
3) Do criminal justice majors perform better than other majors?
4) Does education add incremental validity to cognitive ability?

RESULTS

Is Education a Valid Predictor of Police Performance?

As shown in Table 4.1, the meta-analysis results indicate that education is a valid predictor of all criteria except for commendations and injuries. Better-educated officers perform better in the academy (especially shorter academies), receive higher supervisor evaluations of job performance, have fewer disciplinary problems and accidents, are assaulted less often, use force less often, and miss fewer days of work than their less educated counterparts. From these data, departments appear justified in requiring applicants to have college degrees or a minimum number of college hours.

Meta-analysis results can generalize across studies and situations when the percentage of variance expected by sampling error, measurement error, and range restriction is at least 75% of the observed variance in correlations (Hunter & Schmidt, 1990). For criteria that initially had a high amount of variability, outliers were removed by deleting the top and bottom 5% of the correlations. As can be seen in Table 4.1, removal of outliers reduced variability to an acceptable level for discipline problems but did not affect the mean validity coefficient. Sampling error still did not explain at least 75% of the variance for commendations, and a search for moderators did not reveal any that significantly explained variability. The search for moderators included type of commendation (e.g., citizen, department, awards), type of department, nature of the sample (e.g., age, education), and location. There were not enough studies to properly search for moderators for the injuries and times assaulted criteria.

Table 4.1: Meta-analysis results of the validity of education

Type of Test	K	N	r	95% Confidence Interval		ρ	95% Credibility Interval		Var	Q_w
				Lower	Upper		Lower	Upper		
Academy Grades	32	6,153	.26	.24	.29	.38	.38	.38	100%	19.78
<20 week academy	16	2,477	.31	.27	.35	.46	.46	.46	100%	7.49
>20 week academy	15	3,473	.25	.22	.28	.36	.36	.36	100%	4.68
Supervisor ratings	54	9,120	.17	.12	.21	.28	.16	.40	80%	67.52
Activity	17	4,751	.05	.03	.08	.09	.03	.14	89%	19.09
Commendations	24	6,737	- .03	- .11	.04	- .04	- .30	.21	21%	111.33*
Outliers removed	22	6,427	- .03	- .09	.03	- .04	- .24	.16	29%	74.71*
Discipline problems	54	21,416	- .08	- .10	- .05	- .12	- .21	- .04	73%	73.51*
Outliers removed	51	20,896	- .07	- .09	- .06	- .12	- .17	- .07	89%	55.97
Absenteeism	18	5,669	- .10	- .13	- .07	- .14	- .14	- .14	100%	15.56
Vehicle Accidents	4	1,281	- .17	- .23	- .12	- .23	- .23	- .23	100%	0.81
Injuries	10	3,865	- .06	- .12	.00	- .08	- .25	.09	32%	31.48*
Times assaulted	3	1,399	- .11	- .17	- .04	- .14	- .24	- .04	56%	5.40
Use of force	10	5,217	- .07	- .10	- .04	- .12	- .17	- .06	82%	12.20

K=number of studies, N=sample size, r = mean correlation, ρ = mean correlation corrected for range restriction, criterion unreliability, and predictor reliability, VAR = percentage of variance explained by sampling error and study artifacts, Q_w = the within group heterogeneity

Though a search for moderators was not required in the case of academy grades, it was found that education was a better predictor of grades in shorter academies (r = .31) than in longer academies (r = .25, Z_r = 2.48). The implication of this finding might be that shorter academies are more difficult than longer ones because they contain the same amount of information but teach it in a shorter period of time. More educated cadets are able to handle this increased difficulty more easily than their less formally educated peers.

Though the meta-analysis results support the overall validity of education requirements, it might be fruitful to investigate this issue further. That is, is the relationship between education and performance linear (more education is better than less education) or is the relationship curvilinear such that education beyond a certain point does not enhance performance? Unfortunately, a traditional meta-analysis will not allow us to answer this question. However, we can get insight into this question by combining the raw data from several studies. To do this, I located nine studies that either provided their raw data in an appendix or included frequency tables listing education and performance ratings or academy grades. For each study I first standardized the performance ratings or academy grades using means and standard deviations and then combined the education levels and standardized performance ratings and academy grades into one database.

ANOVAs were then conducted to determine an overall significance for education in the new data set, and Tukey Tests were conducted to determine the significance of differences among the individual means. Consistent with the meta-analysis results, the ANOVAs revealed significant effects for education for both academy grades, F (4, 886) = 11.96, $p < .0001$ and supervisor ratings of performance, F (4, 1926) = 8.17, $p < .0001$. As shown in Table 4.2, the planned comparisons indicated that people with associate's degrees and bachelor's degrees significantly outperformed those with high school diplomas both in the academy and in patrol performance, and people with bachelor's degrees performed significantly better in the academy than those with associate's degrees.

Table 4.2
Standardized Performance Ratings and Academy Grades by Education Level

Education level	Academy Grades		Supervisor Ratings	
	N	Standard Score	N	Standard Score
GED or H.S Diploma	376	- .23[c]	915	- .12[a]
12-64 college hours	100	- .05[ac]	391	.02[b]
Associate's degree	98	.16[ab]	139	.07[bc]
> 64 college hours	62	.01[ac]	45	.29[bc]
Bachelor's degree	255	.32[b]	441	.18[c]

Note: Means in a column that do not share the same superscript are significantly different from one another

Is Education as Valid a Predictor Early in an Officer's Career as Later?

From the meta-analysis results and the analysis described above, it seems clear that officers with college degrees perform better in the academy and on-the-job than those with only a GED or high school diploma. What is not so clear is whether this relationship is true for both new and experienced officers. As alluded to previously, younger officers are most likely in their probationary periods and still learning the job. A possible reason that education might not predict performance as well during this period is that the probationary period is a time of making mistakes and then learning from those mistakes. Thus, it may be that better educated cadets make as many mistakes as less educated cadets, but learn better from these mistakes and subsequently perform better as patrol officers.

To get a better idea of the moderating effect of experience on the relationship between education and performance, five separate data sets were combined that included information on education, patrol performance, and the number of years the officer had been with the department.

That is, would education best predict the performance of officers after they had a chance to get used to the job, complete their field training, and establish their own style? Such a relationship would be consistent with the results of Helmreich, Sawin, and Carsrud (1986) who found that ability best predicted performance during the first two years of work and personality best predicted performance in following years. Helmreich *et al.* labeled the process of different variable predicting early in a career versus later in a career the "Honeymoon Effect." Smith and Aamodt (1996) and Michals and Higgins (1994) reported findings consistent with the idea of a *Honeymoon Effect* in police performance.

To further test the possibility of a *Honeymoon Effect* with education and police performance, the Smith and Aamodt (1996) and Michals and Higgins (1995) data sets were combined with four other data sets in my files. This combined data set contained education, experience, and performance information for 1,003 officers. Because the ratings of overall performance in each dataset were made on different rating scales, the performance ratings were converted to standard scores using the means and standard deviations from each data set.

As shown in Table 4.3, an analysis of variance indicated that overall, officers with bachelor's degrees or higher outperformed lesser-educated officers ($F_{2, 997} = 7.10, p < .0009$). Further analyses indicate that this effect occurred with experienced officers but not for officers in their first two years on the job ($F_{2, 997} = 3.73, p < .02$). There was also a significant main effect for experience ($F_{1, 997} = 6.74, p < .009$).

Does Education Add Incremental Validity to Cognitive Ability Tests?

As you might recall from Chapter 3, the meta-analysis on cognitive ability tests indicated that cognitive ability tests were significant predictors of academy and patrol performance. Given that both cognitive ability and education significantly predict academy and patrol performance, the questions arise about whether the two are tapping the same construct or whether the combination of cognitive ability and education will be more predictive than cognitive ability alone.

To answer this question, two analyses were conducted. The first was designed to use meta-analysis techniques to look at the correlation between education and cognitive ability. Data from 4,776 subjects in 12 studies containing correlations between education and cognitive ability were utilized. From these data, a mean correlation of .29 between education and cognitive ability was observed. The magnitude of this coefficient indicates that though the two are significantly correlated, they do not tap identical constructs.

The second analysis was designed to determine if education added incremental validity to cognitive ability. To do this, the raw data from several studies containing information on both education and cognitive ability were combined into one data set. Because each study used separate measures of cognitive ability, education, and performance, each subject's data were converted into standard scores based on the means and standard deviations from their respective studies. The standardized scores for education and cognitive ability were then entered into a regression to predict the standardized performance ratings and standardized academy grades.

Table 4.3
Interaction of Experience and Education on Patrol Performance

Education level	Experience		Total
	< 2 years	2 or more	
High school	- .09[ab] (0.92) n = 164	- .16[a] (0.98) n =178	- .13[a]
Some college/associate's degree	- .26[a] (1.02) n = 65	.07[b] (0.94) n = 216	- .09[a]
Bachelor's degree	.04[ab] (1.02) n = 290	.34[c] (1.12) n = 90	.19[b]
Total	- .10	.08	.00

Notes
Standard deviations are in parentheses
Means not sharing the same superscript are significantly different from one another

As shown in Table 4.4, education added incremental validity to cognitive ability in predicting both criteria. This analysis provides justification for using a combination of cognitive ability and education to select law enforcement personnel.

Table 4.4
Regression analysis results

Variable	Academy Grades (n=539)			Patrol Performance (n=608)		
	r^2	R^2	$p <$	r^2	R^2	$p <$
Cognitive ability	.327	.327	.001	.163	.163	.001
Education level	.015	.342	.001	.127	.288	.001

Do Criminal Justice Majors Perform Better than Other Majors?

As shown in Table 4.5, there were no significant differences in academy or patrol performance between criminal justice majors and other majors. There were not sufficient data to compare criminal justice majors to related majors (e.g., psychology, sociology) versus non-related majors (e.g. business, art). These meta-analysis results, especially when combined with the Weirman (1978) and Finnigan (1976) studies that could not be included in the meta-analysis because they did not include the proper statistics, seem to suggest that college major is not a major factor in academy or patrol performance.

Discussion

The results of this meta-analysis strongly support providing preference to applicants with college degrees. The results indicate that:

- Education is positively related to academy grades and supervisor ratings of patrol performance and negatively related to absenteeism, injuries, automobile accidents, and tenure.
- Cadets with bachelor's degrees have better grades in the

academy than cadets with associate's degrees or high school diplomas and cadets with associate's degrees have better grades in the academy than cadets with high school diplomas.

- Officers with at least an associate's degree receive higher supervisor ratings of performance than their less educated counterparts.
- The benefits of a college education don't show-up in performance ratings during the first two years of tenure. After two years, however, officers with bachelor's degrees outperform officers with associate's degrees or high school diplomas.
- The relationship between education and academy and patrol performance is not simply a function of intelligence, because education added incremental validity to cognitive ability.
- Criminal justice majors perform at similar levels to other majors.

Though the meta-analysis results conclusively answer several questions about the relationship between education and police performance, further research is necessary in several areas. First, more studies are needed that include both education levels and measures of cognitive ability to get a clearer picture of the extent to which education adds incremental validity to cognitive ability.

Second, more studies are needed looking at differences in performance associated with the various college majors. Though this meta-analysis indicates no real effect for college major, more studies are needed; especially those that include information on cognitive ability. That is, if two officers have equal cognitive ability but different college majors, will their performance differ?

Third, research is necessary to investigate the relationship between high school and college GPA and academy and patrol performance. This area is especially interesting in light of the recent meta-analysis by Roth, BeVier, Switzer, and Schippmann (1996) indicating that both high school and college grades significantly predict performance in a variety of jobs. It would seem intuitive that high school and college grades would be highly related to academy performance, but to date, only one study was found addressing this issue. In that study (Griffin, 1980), high school GPA was not significantly correlated with patrol performance (r = .01).

Fourth, more studies are needed to investigate the "honeymoon effect" that was postulated between education and patrol performance. Such studies would focus on whether tenure moderates the relationship between education and performance.

Fifth, research is needed to determine if officers who earn their degrees after being hired differ from those who had their degrees prior to being hired. Furthermore, did the officers' patrol performance improve while they were taking college courses, after their course work was completed, or not at all?

Sixth and finally, the relationship between education and performance in command positions is in dire need of investigation. Given that education relates to performance at the academy and patrol levels, it makes sense that the relationship would be even stronger for such policy makers as lieutenants, captains, and chief. Future research should seek to confirm or deny such an assumption.

Table 4.5
Meta-analysis results of the validity of having a criminal justice major

Type of Test	K	N	r	95% Confidence Interval		ρ	95% Credibility Interval		Var	Q_w
				Lower	Upper		Lower	Upper		
Academy Grades	6	976	.01	- .05	.08	.02	- .07	.11	80%	7.48
Supervisor ratings	9	1,706	- .02	- .09	.05	- .03	- .23	.17	49%	18.20*
Commendations	4	1,158	.05	- .01	.11	.06	- .09	.22	40%	10.03*
Discipline problems	10	2,621	.00	- .04	.04	.00	.00	.00	100%	0.00
Absenteeism	1	129	- .11							
Vehicle accidents	1	46	- .10							

K=number of studies, N=sample size, r = mean correlation, ρ = mean correlation corrected for range restriction, criterion unreliability, and predictor reliability, VAR = percentage of variance explained by sampling error and study artifacts, Q_w = the within group heterogeneity

Chapter References

*Aamodt, M. G., & Flink, W. (2001). Relationship between educational level and cadet performance in a police academy. *Applied H.R.M. Research, 6*(1), 75-76.

*Abraham, J. D., & Morrison, J. D. (2003). Relationship between the Performance Perspectives Inventory's Conscientiousness scale and job performance of corporate security guards. *Applied H.R.M. Research*, 8(1), 45-48.

*Agyapong, O. A. (1988). *The effect of professionalism on police job performance: An empirical assessment.* Unpublished doctoral dissertation, Florida State University.

Arthur, W., Bennett, W., & Huffcutt, A. I. (2001). *Conducting meta-analysis using SAS.* Mahwah, NJ: Erlbaum.

*Baratta, C. R. (1998). *The relationship between education and police work performance.* Unpublished master's thesis, University of Baltimore.

*Barbas, C. (1992). *A study to predict the performance of cadets in a police academy using a modified cloze reading test, a civil service aptitude test, and educational level.* Unpublished doctoral dissertation, Boston University.

*Bennett, R. R. (1978). The effects of education on police values and performance: A multivariate analysis of an explanatory model. In Wellford, C. (Ed.) *Quantitative studies in criminology.* Beverly Hills, CA: Sage Publications.

*Boes, J. O., Chandler, C. J., & Timm, H. W. (1997). *Police integrity: use of personality measures to identify corruption-prone officers.* Monterey, CA: Defense Personnel Security Research Center.

*Boyce, T. N. (1988). *Psychological screening for high-risk police specialization.* Unpublished doctoral dissertation, Georgia State University.

*Brewster, J., & Stoloff, M. (2003). Relationship between IQ and first-year performance as a police officer. *Applied H.R.M. Research, 8*(1), 49-50.

Buracker, C. D. (1977). The educated police officer: Asset or liability. *The Police Chief*, August, 90-94.

*Buttolph, S. E. (1999). *Effect of college education on police behavior: Analysis of complaints and commendations.* Unpublished master's thesis, East Tennessee State University.

*Campa, E. E. (1993). *The relationship of reading comprehension and educational achievement levels to academy and field training performance of police cadets.* Unpublished doctoral dissertation, Texas A&M University.

Carter, D. L., & Sapp, A. D. (1992). College education and policing: Coming of age. *FBI Law Enforcement Bulletin, 61*(1), 8-14.

*Cascio, W. F. (1977). Formal education and police officer performance. *Journal of Police Science and Administration*, *5*(1), 89-96.

Castro v. Beecher, 451 F. 2d 724 (C.A.1, 1972).

*Champion, D. H. (1994). *A study of the relationship between critical thinking levels and job performance of police officers in a medium size police department in North Carolina.* Unpublished doctoral dissertation, North Carolina State University.

Chicano Police Officer's Association et al., Appellants v. Robert V. Stover et al., 526 F.2nd 431, 1976

*Cohen, B. & Chaiken, J. M. (1973). *Police Background Characteristics and Performance.* Lexington, MA: Lexington Books.

*Copley, W. H. (1987). *Using education, academy, and field training scores to predict success in a Colorado police department.* Unpublished doctoral dissertation, Colorado State University.

*Dailey, J. D. (2002). *An investigation of police officer background and performance: An analytical study of the effect of age, time in service, prior military service, and educational level on commendations.* Unpublished doctoral dissertation, Sam Houston State University.

*Daley, R. E. (1978). *The relationship of personality variables to suitability for police work.* Unpublished doctoral dissertation, Florida Institute of Technology.

Daniel, E. D. (1982). The effect of a college degree on police absenteeism. *The Police Chief, 49*(9), 70-71.

*Davis, R., & Rostow, C. (2003). Relationship between cognitive ability and background variables and disciplinary problems in law enforcement. *Applied H.R.M. Research, 8*(2), 77-80.

Davis v. Dallas, 777 F.2d 205 (5th Cir. 1985).

*Decker, L. K., & Huckabee, R. G. (2002). Raising the age and education requirements for police officers: Will too many women and minority candidates be excluded? *Policing: An International Journal of Police Strategies & Management, 25*(4), 789-802.

*Del Castillo, V. (1984). *Education and the police: A study of the relationship between higher education and police officer performance*. Unpublished master's thesis, John Jay College.

*Dibb, G. S. (1978). *A cross-validated comparison of models for the prediction of academy performance and job tenure of police officer recruits*. Unpublished doctoral dissertation, University of Hawaii.

*Dorner, K. R. (1991). *Personality characteristics and demographic variables as predictors of job performance in female traffic officers*. Unpublished doctoral dissertation, United States International University.

*Dorsey, R. R. (1994). *Higher education for police officers: An analysis of the relationships among higher education, belief systems, job performance, and cultural awareness*. Unpublished doctoral dissertation, University of Mississippi.

*Duignan, J. F. (1978). Education's role in the quest for professionalism. *The Police Chief, 45*(8), 29.

Feldman, K. A., & Newcomb, T. M. (1969). *The impact of college on students, Vol 1: An analysis of four decades of research*. San Francisco: Jossey-Bass.

*Ferrell, N. K. (1994). *Police officers' receptivity to community policing*. Unpublished doctoral dissertation, East Texas State University.

Finckenauer, J. O. (1975). Higher education and police discretion. *Journal of Police Science and Administration, 3*(4), 450-457.

Finnigan, J. C. (1976). A study of the relationships between college education and police performance in Baltimore, Maryland. *The Police Chief, 43*(8), 60-62.

*Gardner, J. F. (1994). *The predictive validity of psychological testing in law enforcement*. Unpublished master's thesis, University of Alabama.

Geary, D. P. (1970). College educated cops - three years later. *The Police Chief, 37*(8), 59-62.

*Geary, D. P. (1979). *A study of the relationship of selected educational factors to police performance*. Unpublished doctoral dissertation, University of Nevada at Reno.

Genz, J. L., & Lester, D. (1977). Military service, education, and authoritarian attitudes of municipal police officers. *Psychological Reports, 40*, 402.

*Geraghty, M. F. (1986). *The California Personality Inventory test as a predictor of law enforcement officer job performance*.

Unpublished doctoral dissertation, Florida Institute of Technology.

Goldstein, H. (1977). *Policing a free society.* Cambridge, MA: Ballinger.

*Gottlieb, M. C., & Baker, C. F. (1974). Predicting police officer effectiveness. *The Journal of Forensic Psychology, 6*, 35-46.

*Graziano, J. R. (1995). *The relationship between police officers' level of education and work performance.* Unpublished doctoral dissertation, Southern Illinois University at Carbondale.

*Griffin, G. R. (1980). *A study of the relationships between levels of college education and police patrolmen's performance.* Saratoga, CA: Century Twenty One Publishing.

*Griffiths, R. F., & McDaniel, Q. P. (1993). Predictors of police assaults. *Journal of Police and Criminal Psychology, 9*(1), 5-9.

Gross, S. (1973). Higher education and police: Is there a need for a closer look? Journal of *Police Science and Administration, 4*(1), 477.

*Hamack, R. F. (1988). *Pre-academy placement in the Washington State Patrol: Factors associated with academy and job performance.* Unpublished master's thesis, Central Washington University.

*Hankey, R. O. (1968). *Personality correlates in a role of authority: The police.* Unpublished doctoral dissertation, University of Southern California.

*Hankey, R. O., Morman, R. R., Kennedy, P., & Heywood, H. L. (1965). TAV selection system and state traffic officer job performance. *Police*, March-April, 10-13.

Helmreich, R. L., Sawin, L. L., & Carsrud, A. L. (1986). The honeymoon effect in job performance: Temporal increases in the predictive power of achievement motivation. *Journal of Applied Psychology, 71*, 185-188.

*Helrich, K. L. (1985). *The use of hardiness and other stress-resistance resources to predict symptoms and performance in police academy trainees.* Unpublished doctoral dissertation, California School of Professional Psychology, San Diego.

*Heyer, T. (1998). *A follow-up study of the prediction of police officer performance on psychological evaluation variables.* Unpublished doctoral dissertation, Minnesota School of Professional Psychology.

*Hooper, M. K. (1988). *Relationship of college education to police officer job performance.* Unpublished doctoral dissertation, Claremont Graduate School.

Hunter, J. E., & Schmidt, F. L. (1990). Methods of meta-analysis: Correcting error and bias in research findings. Newbury Park, CA: Sage Publications.

Ice, et al. v. Arlington County, et al., No. 76-2194, 4th Circuit, May 16, 1977.

*Jayaraj, E. A. S. (1984). *A predictive validity study of police officer selection.* Unpublished master's thesis, Southern Connecticut State University.

*Johnson, T. A. (1998). *The effects of higher education/military service on achievement levels of police academy cadets.* Unpublished doctoral dissertation, Texas Southern University.

Kakar, S. (2003). Race and police officers' perceptions of their job performance: An analysis of the relationship between police officers' race, educational level, and job performance. *Journal of Police and Criminal Psychology, 18*(1), 45-56.

Kakar, S. (1998). Self-evaluations of police performance: An analysis of the relationship between police officers' education level and job performance. *Policing: An International Journal of Police Strategies and Management, 21*(4), 632-647.

*Kappeler, V. E., Sapp, A. D., & Carter, D. L. (1992). Police officer higher education, citizen complaints, and departmental rule violations. *American Journal of Police, 11*(2), 37-54.

*Kayode, O. (1973). *Predicting performance on the basis of social background characteristics: The case of the Philadelphia Police Department.* Unpublished doctoral dissertation, University of Pennsylvania.

*Kedia, P. R. (1985). *Assessing the effect of college education on police job performance.* Unpublished doctoral dissertation, University of Southern Mississippi.

*Krimmel, J. T. (1996). The performance of college-educated police: A study of self-rated police performance measures. *America Journal of Police, 15*(1), 85-95.

*Lester, D. (1979). Predictors of graduation from a police training academy. *Psychological Reports, 44*, 362.

*Levy, R. J. (1967). Predicting police failures. *Journal of Criminal Law, Criminology, and Police Science, 58*(2), 265-276.

*Madden, B. L. (1983). *The police and higher education: A study of the relationship between higher education and police officer performance.* Unpublished master's thesis, University of Louisville.

*Madell, J. D., & Washburn, P. V. (1978). Which college major is best for the street cop? *The Police Chief, 45*(8), 40-42.

*Matyas, G. S. (1980). *The relationship of MMPI and background data to police performance.* Unpublished doctoral dissertation, University of Missouri - Columbia.

*McConnell, W. A. (1967). *Relationship of personal history to success as a police patrolman.* Unpublished doctoral dissertation, Colorado State University.

*McGreevy, T. J. (1964). *A field study of the relationship between the formal education levels of 556 officers in St. Louis, Missouri, and their patrol duty performance records.* Unpublished master's thesis, Michigan State University.

*Mealia, R. M. (1990). *Background factors and police performance.* Unpublished doctoral dissertation.

*Michals, J. E., & Higgins, J. M. (1994). *Effects of education level on performance of campus police officers.* Paper presented at the annual Graduate Student Conference in Industrial/Organizational Psychology and Organizational Behavior, Chicago, Illinois.

Miller, J., & Fry, L. J. (1978). Some evidence on the impact of higher education for law enforcement personnel. *The Police Chief,* August, 30-33.

*Morman, R. R., Hankey, R. O., Kennedy, P. K., & Haywood, H. L. (1965). Predicting state traffic officer performance with TAV selection system theoretical scoring keys, *Police,* May-June, 70-73.

*Mullins, W. C., & McMains, M. (1996). Predicting patrol officer performance from a psychological assessment battery: A predictive study. *Journal of Police and Criminal Psychology, 10*(4), 15-25.

*Murrell, D. B. (1982). The influence of education on police work performance. Unpublished doctoral dissertation, Florida State University.

*Palombo, B. J. (1995). *Academic professionalism in law enforcement.* New York: Garland Publishing.

*Patterson, G. T. (2002). Predicting the effects of military service experience on stressful occupational events in police officers. *Policing: An International Journal of Police Strategies & Management, 25*(3), 602-618.

*Peterson, D. S. (2001). *The relationship between educational attainment and police performance.* Unpublished doctoral dissertation, Illinois State University.

*Pibulniyom, S. (1984). *A quantitative analysis of dynamic performance measurements of a southern police department.* Unpublished doctoral dissertation, University of Mississippi.

*Plummer, K. O. (1979). *Pre-employment factors that determine success in the police academy.* Unpublished doctoral dissertation, Claremont Graduate College.

*Poland, J. M. (1976). *An exploratory analysis of the relationship between social background factors and performance criteria in the Michigan State Police.* Unpublished doctoral dissertation, Michigan State University.

Police Executive Research Forum (1989). *The state of police education: Policy direction for the 21st century.* Washington, D.C.: PERF.

President's Commission on Law Enforcement and Administration of Justice (1967). *The challenge of crime in a free society.* Washington, D.C.: U.S. Government Printing Office.

*Quarles, C. L. (1984). *A correlation of police productivity with educational level, age, and seniority of officers in a southern police department.* Paper presented at the annual meeting of the Society for Police and Criminal Psychology, Little Rock, Arkansas.

Reming, G. C. (1988). Personality characteristics of supercops and habitual criminals. *Journal of Police Science and Administration, 16*(3), 163-167.

*Roberg, R. R. (1978). An analysis of the relationships among higher education, belief systems, and job performance of patrol officers. *Journal of Police Science and Administration, 6*, 336-344.

*Roberts, J. (1984). *The relationship of higher education in Oklahoma Highway Patrol troopers' performance.* Unpublished doctoral dissertation, University of Oklahoma.

*Rose, J. E. (1995). *Consolidation of law enforcement basic training academies: An evaluation of pilot projects.* Unpublished doctoral dissertation, Northern Arizona University.

Roth, P. L., BeVier, C. A., Switzer, F. S., & Schippmann, J. S. (1996). Meta-analyzing the relationship between grades and job performance. *Journal of Applied Psychology, 81*(5), 548-556.

*Sanderson, B. E. (1977). Police officers: The relationship of college education to performance. *The Police Chief, 44*(8), 62-63.

*Scarfo, S. (2002). Relationship between police academy performance and level of education. *Applied H.R.M. Research, 7*(1), 24.

*Schaller, G. R. (1990). *Use of the Armed Services Vocational Aptitude Battery to predict training outcomes in female military police trainees.* Unpublished doctoral dissertation, Auburn University.

Scott, W. R. (1986). College education requirements for police entry level and promotion: A study. *Journal of Police and Criminal Psychology, 2*(1), 10-28.

*Shaver, D. P. (1980). *A descriptive study of police officers in selected towns of northwest Arkansas.* Unpublished doctoral dissertation, University of Arkansas.

*Sherman, L. W., & Blumberg, M. (1981). Higher education and police use of deadly force. *Journal of Criminal Justice,* 9, 317-331.

Sherwood, C. W. (1994). *The relationship between higher education and job satisfaction: A study of municipal police officers in two cities.* Unpublished doctoral dissertation, University of New Haven (CT).

Smith, A. B., Locke, B., & Fenster, A. (1970). Authoritarianism in policemen who are college graduates and non-college oriented police. *Journal of Criminal Law, Criminology, and Police Science, 50,* 440-443.

*Smith, D. C., & Ostrom, E. (1974). The effects of training and education on police attitudes and performance: A preliminary study. In Jacob, Herbert (Ed.). *The Potential for Reform of Criminal Justice.* Beverly Hills, CA: Sage Publications.

*Smith, S. M., & Aamodt, M. G. (1997). Relationship between education, experience, and police performance. *Journal of Police and Criminal Psychology, 12*(2), 7-14.

*Staff, T. G. (1992). *The utility of biographical data in predicting job performance: Implications for the selection of police officers.* Unpublished doctoral dissertation, University of Toledo.

*Stafford, A. R. (1983). *The relationship of job performance to personal characteristics of police patrol officers in selected Mississippi police departments.* Unpublished doctoral dissertation, University of Southern Mississippi.

*Sterne, D. M. (1960). Use of the Kuder Preference Record, Personal, with police officers. *Journal of Applied Psychology, 44*(5), 323-324.

*Sterrett, M. R. (1984). *The utility of the Bipolar Psychological Inventory for predicting tenure of law enforcement officers.* Unpublished doctoral dissertation, Claremont Graduate College.

*Talley, J. E., & Hinz, L. D. (1990). *Performance prediction of publis safety and law enforcement personnel.* Springfield, IL: Charles C. Thomas.

*Tidwell, H. D. (1993). *The predictive value of biographical data: An analysis of using biodata to predict short tenure or unsuitability of police officers.* Unpublished doctoral dissertation, University of Texas-Arlington.

*Tompkins, L. P. (1977). *An evaluation of police academy training upon selected recruits and its relationship to job performance.* Unpublished master's thesis, Rollins College, FL.

*Topp, B. W., & Kardash, C. A. (1986). Personality, achievement, and attrition: Validation in a multiple-jurisdiction police academy. *Journal of Police Science and Administration, 14*(3), 234-241.

Trojanowicz, R. C., & Nicholson, T. G. (1976). A comparison of behavioral styles of college graduate police officers. *The Police Chief, 43*(8), 56-59.

*Truxillo, D. M., Bennett, S. R., & Collins, M. L. (1998). College education and police job performance: A ten-year study. *Public Personnel Management, 27*(2), 269-280.

Tucker, M. L., & Hyder, A. K. (1978). Some practical considerations in law enforcement education. *The Police Chief, 45*(8), 26-28.

*Uno, E. A. (1979). *The prediction of job failure: A study of police officers using the MMPI.* Unpublished doctoral dissertation, California School of professional Psychology, Berkeley.

*Varela, J. G., Scogin, F. R., & Vipperman, R. K. (1999). Development and preliminary validation of a semi-structured interview for the screening of law enforcement candidates. *Behavioral Science and the Law, 17*(4), 467-481.

Walker, R. B. (1994). Management implications of education standards. *The Police Chief, 61*(11), 24-26.

*Ward, J. C. (1981). *The predictive validity of personality and demographic variables in the selection of law enforcement officers.* Unpublished doctoral dissertation, University of South Florida.

*Waugh, L. (1996). *Police officer recruit selection: Predictors of academy performance.* Queensland, Australia: Queensland Police Academy.

*Weirman, C. L. (1978). Variances of ability measurement scores obtained by college and non-college educated troopers. *The Police Chief, 45*(8), 34-36.

*Wellman, R. J. (1982). *Accident proneness in police officers: Personality factors and problem drinking as predictors of injury claims of sate troopers.* Unpublished doctoral dissertation, University of Connecticut.

*Wexler, N., & Sullivan, S. M. (1982). *Concurrent validation of a prototype selection test for entry-level police officers.* Trenton, NJ: New Jersey Department of Civil Service, Division of Examinations.

*Wiens, A., Purintun, C., & Connelly, M. (1997). *Factors associated with successful completion of the Oklahoma Highway Patrol Academy.* Oklahoma City, OK: Oklahoma Criminal Justice Resource Center.

*Wilson, H. T. (1994). *Post-secondary education and the police officer: A study of the effect on the frequency of citizens' complaints.* Unpublished doctoral dissertation, Golden Gate University.

*Wolff, T. E. (1991). *The relationship between a college education and police performance.* Unpublished master's thesis, University of South Florida.

*Wolfskil, J. R. (1989). *Higher education and police performance.* Unpublished doctoral dissertation, University of Kansas.

*Woods, D. A. (1991). *An analysis of peace officer licensing revocations in Texas.* Unpublished doctoral dissertation, Sam Houston State University.

Worden, R. E. (1990). A badge and a baccalaureate: Policies, hypotheses, and further evidence. *Justice Quarterly*, 7(3), 565-592.

*Wymer, C. W. (1996). *A comparison of the relationships between level of education, job performance, and beliefs on professionalism within the Virginia State Police.* Unpublished doctoral dissertation, Virginia Polytechnic Institute and State University.

* Indicates that the reference was used in the meta-analysis

Chapter 5
Previous Military Service and Police Performance

Giving preference to applicants with prior military experience is a common practice with law-enforcement agencies. There are two main ideas behind providing such a preference: rewarding veterans for their service to the country and assumed validity.

In rewarding veterans for prior military service, an agency is not making any assumptions about validity. Instead, it provides some form of preferential hiring as a "thank you" for serving the country. Usually this "thank you" is mandated by state or local law and comes in the form of adding 5 or 10 points to a veteran's civil service examination score. On Federal civil service exams, veterans receive 5 points added to their examination scores.

At times, this "thank you" is done more informally when an agency decides to "break any ties" between applicants by opting for military veterans. The informal granting of such a preference is common in a selection system using something such as a "rule of three" or "rule of five" wherein the names of the top three or five scores on an exam are given to the Chief. The chief then has sole authority to choose which of the top three or five scorers he or she wants to hire.

From a legal perspective, there is nothing wrong with providing preference for military veterans because the 1964 Civil Rights Act exempts such preference from adverse impact violations (examples of other exemptions include valid tests, bona fide seniority systems, and national security).

A second common reason for giving preference to veterans is the assumption that military veterans will be better cops than non-veterans because the military environment and training techniques are similar to those in many law enforcement agencies. It is the purpose of this chapter to explore the validity of this assumption by conducting a meta-analysis of research on this topic.

Results

On the basis of the available studies on this issue, there is little empirical support for the notion that officers with military experience are more successful than those without such experience. As shown in Table 5.1, the only significant difference between officers with military experience and those without is that officers with military experience were slightly more likely to have received commendations ($r = .07$, $\rho = .10$) than were officers without military experience. As also shown in Table 5.1, with the exception of performance ratings, sampling error and artifacts explain at least 75% of the variation in validity across studies, thus these results can be generalized across departments.

Because sampling error explained less than 75% of the variance for supervisor ratings of patrol performance, a search for moderators was conducted. These potential moderators included department type, (e.g., police, sheriff), location (urban, rural), region (south, west, east, midwest), study date, and study source (journal article, dissertation). Unfortunately, no moderator could be found that significantly reduced the variability among studies. The Boyce (1988) study deserves special notice because it involved a specialized sample (e.g., SWAT, narcotics, vice). The finding that previous military experience was positively related to performance in this study ($r = .26$), suggests the need for further research as military experience may be useful for specialized assignments but not regular patrol assignments.

Conclusion

The general lack of a relationship between prior military experience and most measures of police performance is an important finding because giving preference for veteran status will result in adverse impact against females. Though such adverse impact is not illegal due to the aforementioned exemption, it is counter to the desire of most law enforcement agencies to increase the number of females in law enforcement positions. Furthermore, this adverse impact would not be accompanied by a higher quality workforce.

The meta-analysis results suggest that in general, officers with prior military experience will receive more commendations and perhaps be assaulted less often (only 2 studies) than their non-military counterparts but will not perform better on other criteria. Two areas appear ripe for further study. One study (Boyce, 1988) found a significant positive relationship between military experience and performance by SWAT personnel and two studies (Hooper, 1988; Stohr-Gillmore, Stohr-Gillmore, & Kistler, 1990) found a positive relationship between military experience and field training experience. Both of these findings make sense but more studies are needed to confirm these relationships.

Table 5.1
Meta-analysis results for the validity of military experience

Criterion	K	N	r	95% Confidence Interval		ρ	90% Credibility Interval		Var	Q_w
				Lower	Upper		Lower	Upper		
Academy Grades	9	2,061	.02	-.02	.07	.04	-.02	.09	90%	9.95
Field training	2	165	.15	-.01	.30	.20	.20	.20	100%	0.14
Supervisor ratings	16	4,090	-.03	-.11	.05	-.05	-.37	.28	23%	71.15*
Commendations	8	1,337	-.07	.02	.13	.10	.02	.18	81%	9.80
Discipline problems	14	6,296	-.02	-.05	.00	-.04	-.04	-.04	100%	7.21
Sick leave	5	1,109	-.01	-.07	.05	-.02	-.02	-.02	100%	4.71
Times assaulted	2	320	-.08	-.19	.03	-.11	-.11	-.11	100%	0.31
Use of force	4	2,067	-.03	-.07	.01	-.05	-.05	-.05	100%	1.87

K=number of studies, N=sample size, r = mean correlation, ρ = mean correlation corrected for range restriction, criterion unreliability, and predictor reliability, VAR = percentage of variance explained by sampling error and study artifacts, Q_w = the within group heterogeneity

References

*Aadland, R. L. (1981). *The prediction of use of deadly force by police officers in simulated field situations.* Unpublished doctoral dissertation, California School of Professional Psychology, Los Angeles.

*Azen, S. P., Snibbe, H. M., Montgomery, H. R., Fabricatore, J. , & Earle, H. (1974). A longitudinal predictive study of success and performance of law enforcement officers. *American Journal of Community Psychology, 1*(2), 79-86.

*Boyce, T. N. (1988). *Psychological screening for high-risk police specialization.* Unpublished doctoral dissertation, Georgia State University.

*Davis, R., & Rostow, C. (2003). Relationship between cognitive ability and background variables and disciplinary problems in law enforcement. *Applied H.R.M. Research, 8*(2), 77-80.

*Gonder, M. L. (1998). *Personality profiles of police officers: Differences in those that complete and fail to complete a police training academy.* Unpublished master's thesis, University of North Carolina-Charlotte.

*Gottlieb, M.C., & Baker, C.F. (1974). Predicting police officer effectiveness. *The Journal of Forensic Psychology, 6*, 35-46.

*Griffiths, R. F., & McDaniel, Q. P. (1993). Predictors of police assaults. *Journal of Police and Criminal Psychology, 9*(1), 5-9.

*Hankey, R. O. (1968). *Personality correlates in a role of authority: The police.* Unpublished doctoral dissertation, University of Southern California.

*Hooper, M.K. (1988). *Relationship of college education to police officer job performance.* Unpublished doctoral dissertation, Claremont Graduate School.

*Kayode, O. (1973). *Predicting performance on the basis of social background characteristics: The case of the Philadelphia Police Department.* Unpublished doctoral dissertation, University of Pennsylvania.

*Kedia, P.R. (1985). *Assessing the effect of college education of police job performance.* Unpublished doctoral dissertation, University of Southern Mississippi.

*Lester, D. (1979). Predictors of graduation from a police training academy. *Psychological Reports, 44*, 362.

*Matyas, G.S. (1980). *The relationship of MMPI and biographical data to police performance.* Unpublished doctoral dissertation. University of Missouri, Columbia.

*McConnell, W.A. (1967). *Relationship of personal history to success as a police patrolman.* Unpublished doctoral dissertation. Colorado State University.

*Melia, R.M. (1990). *Background factors and police performance.* Unpublished doctoral dissertation, State University of New York, Albany.

*Murrell, D. B. (1982). *The influence of education on police work performance.* Unpublished doctoral dissertation, Florida State University.

*Patterson, G. T. (2002). Predicting the effects of military service experience on stressful occupational events in police officers. *Policing: An International Journal of Police Strategies & Management, 25*(3), 602-618.

*Plummer, K. O. (1979). *Pre-employment factors that determine success in the police academy.* Unpublished doctoral dissertation, Claremont Graduate College.

*Poland, J. M. (1976). *An exploratory analysis of the relationship between social background factors and performance criteria in the Michigan State Police.* Unpublished doctoral dissertation, Michigan State University

*Rose, J.E. (1995). *Consolidation of law enforcement basic training academies: An evaluation of pilot projects.* Unpublished doctoral dissertation. Northern Arizona University.

*Scarfo, S. J. (2002). Relationship between police academy performance and cadet level of education and cognitive ability, *Applied H.R.M. Research, 7*(1), 24.

*Shaver, D. P. (1980). *A descriptive study of police officers in selected towns of northwest Arkansas.* Unpublished doctoral dissertation, University of Arkansas.

*Shusman, E. J., Inwald, R. E., & Landa, B. (1984). Correction officer job performance as predicted by the IPI and MMPI. *Criminal Justice and Behavior, 11*(3), 309-329.

*Stohr-Gillmore, M. K., Stohr-Gillmore, M. W., & Kistler, N. (1990). Improving selection outcomes with the use of situational interviews: Empirical evidence from a study of correctional officers for new generation jails. *Review of Public Personnel Administration, 10*(2), 1-18.

*Talley, J. E., & Hinz, L. D. (1990). *Performance prediction of public safety and* law enforcement personnel: A study in race and gender differences and MMPI subscales. Springfield, IL: Charles C. Thomas.

*Uno, E. A. (1979). *The prediction of job failure: A study of police officers using the MMPI.* Unpublished doctoral dissertation, California School of Professional Psychology, Berkeley.

*Wiens, A., Purintun, C., & Connelly, M. (1997). *Factors associated with successful completion of the Oklahoma Highway Patrol Academy.* Oklahoma City, OK: Oklahoma Criminal Justice Resource Center.

* Indicates study was used in the meta-analysis

Chapter 6
Background Information

A common practice in law enforcement selection is to conduct a background investigation on applicants prior to their being hired. The purpose of the investigation is to determine if there are factors in an applicant's background that would cast doubt on his or her ability to perform well in a law enforcement context. Such factors include driving record, arrest record, credit history, and discipline problems in the military, in school, and at work. This chapter will describe the meta-analysis results of the limited research on background information. Although some studies included information on such background factors as marital status and number of children, these factors were not included in this analysis because using them to make an employment decision would be illegal.

Meta-Analysis Results

There were relatively few studies investigating the relationship between background problems and police performance. Perhaps the reason for this lack of research is that officers who had arrest records and poor work histories were not hired and thus their law enforcement performance could not be studied. With that said, as shown in Tables 6.1 and 6.2, officers with arrest records and a history of problems at work are less likely to perform well than their counterparts without legal and work problems. A higher number of traffic tickets received prior to hire corresponds to lower supervisor ratings, more commendations, and more sick time used. The number of studies on which these conclusions are based is small so more research is still necessary. Although the positive relationship between traffic tickets and commendations seems odd, it may be that whatever personality characteristic is related to getting tickets (e.g., risk taking behavior) is also related to engaging in patrol activities that result in commendations.

Though only based on three studies, the background investigation in general seems to be an excellent predictor of police performance with a mean validity of .27 and a corrected validity of .41. The corrected validity perhaps could be even higher but there was no information available about the reliability of background investigations thus the mean validity could not be corrected for predictor unreliability.

Table 6.1
Meta-analysis results for the validity of applicant background problems

Background variable/Criterion	K	N	r	95% Confidence Interval		ρ	90% Credibility Interval		Var	Q_w
				Lower	Upper		Lower	Upper		
Traffic Tickets										
Academy Grades	2	464	-.01	-.12	.11	-.01	-.28	.26	26%	7.67*
Supervisory Ratings	9	2,425	-.12	-.18	-.07	-.20	-.35	-.06	65%	13.89*
Commendations	3	1,799	.09	.05	.14	.13	.13	.13	100%	0.62
Discipline problems	4	3,530	.04	.00	.07	.06	-.02	.11	64%	6.18*
Sick leave	3	284	.13	.01	.24	.17	.17	.17	100%	1.38
Activity	2	332	-.06	-.16	.05	-.09	-.09	-.09	100%	0.85
Accidents	2	1,837	.02	-.02	.07	-.07	-.07	.14	33%	6.09*
Background rating	3	2,146	.27	.19	.34	.41	.20	.61	47%	6.42*

K=number of studies, N=sample size, r = mean correlation, ρ = mean correlation corrected for range restriction, criterion unreliability, and predictor reliability, VAR = percentage of variance explained by sampling error and study artifacts, Q_w = the within group heterogeneity

Table 6.2
Meta-analysis results for the validity of applicant background problems (cont.)

Background variable/Criterion	K	N	r	95% Confidence Interval		ρ	90% Credibility Interval		Var	Q_w
				Lower	Upper		Lower	Upper		
Arrest Record										
Academy Grades	1	301	-.27							
Supervisory Ratings	9	2,924	-.25	-.34	-.17	-.42	-.71	-.13	42%	21.49*
Discipline Problems	3	2,227	.03	-.03	.09	.05	-.11	.21	28%	10.71*
Problems at Work										
Academy Grades	2	457	-.20	-.29	-.11	-.29	-.29	-.29	100%	0.14
Supervisor Ratings	10	2,178	-.21	-.25	-.16	-.34	-.34	-.34	100%	9.35
Discipline problems	5	4,403	.07	.04	.11	.12	.19	.19	55%	8.88*
Commendations	1	160	-.02							
Activity	1	301	-.12							

K=number of studies, N=sample size, r = mean correlation, ρ = mean correlation corrected for range restriction, criterion unreliability, and predictor reliability, VAR = percentage of variance explained by sampling error and study artifacts, Q_w = the within group heterogeneity

Chapter References

*Boes, J. O., Chandler, C. J., & Timm, H. W. (1997). *Police integrity: Use of personality measures to identify corruption-prone officers.* Monterey, CA: Defense Personnel Security Research Center.

*Cohen, B., & Chaiken, J. M. (1973). *Police background characteristics and performance.* Lexington, MA: Lexington Books.

*Davis, R., & Rostow, C. (2003). Relationship between cognitive ability and background variables and disciplinary problems in law enforcement. *Applied H.R.M. Research, 8*(2), 77-80.

*Dibb, G. S. (1978) *A cross-validated comparison of models for the prediction of academy performance and job tenure of police officer recruits.* Unpublished doctoral dissertation, University of Hawaii, Honolulu.

*Inwald, R. E., & Brockwell, A. L. (1991). Predicting the performance of government security personnel with the IPI and MMPI. *Journal of Personality Assessment, 56*(3), 522-535.

*Inwald, R. E., & Shusman, E. J. (1984). The IPI and MMPI as predictors of academy performance for police recruits. *Journal of Police Science and Administration, 12*(1), 1-11.

*Kauder, B. S., & Thomas, J. C. (2003). Relationship between MMPI-2 and IPI scores and ratings of police officer probationary performance. *Applied H.R.M. Research, 8*(2), in press.

*Matyas, G. S. (1980). *The relationship of MMPI and biographical data to police performance.* Unpublished doctoral dissertation, University of Missouri, Columbia.

*McConnell, W. A. (1967). *Relationship of personal history to success as a police patrolman.* Unpublished doctoral dissertation, Colorado State University.

*Mealia, R. M. (1990). *Background factors and police performance*. Unpublished doctoral dissertation, State University of New York, Albany.

*Palmatier, J. J. (1996). *The big-five factors and hostility in the MMPI and IPI: Predictors of Michigan State Trooper job performance*. Unpublished doctoral dissertation, Michigan State University.

*Poland, J, M. (1976). *An exploratory analysis of the relationship between social background factors and performance criteria in the Michigan State Police*. Unpublished doctoral dissertation, Michigan State University.

*Shaffer, A. M. (1996). *Predictive and discriminative validity of various police officer selection criteria*. Unpublished master's thesis, University of California, Irvine.

*Shaver, D. P. (1980). *A descriptive study of police officers in selected towns of northwest Arkansas*. Unpublished doctoral dissertation, University of Arkansas.

*Staff, T. G. (1992). *The utility of biographical data in predicting job performance: Implications for the selection of police officers*. Unpublished doctoral dissertation, University of Toledo.

*Surrette, M. A., Aamodt, M. G., & Serafino, G. (1990). *Validity of the New Mexico Police Selection Battery.* Paper presented at the annual meeting of the Society for Police and Criminal Psychology, Albuquerque, NM.

*Varela, J. G., Scogin, F. R., & Vipperman, R. K. (1999). Development and preliminary validation of a semi-structured interview for screening law enforcement candidates. *Behavioral Science and the Law, 17*(4), 467-481.

* Study was included in the meta-analysis

Chapter 7
Personality Inventories

Personality inventories are commonly used methods to predict performance in law enforcement settings. Although there are hundreds of personality inventories available, they generally fall into one of two categories based on their intended purpose: measures of psychopathology and measures of normal personality.

Measures of Psychopathology

Measures of psychopathology (abnormal behavior) determine if individuals have serious psychological problems such as depression, bipolar disorder, and schizophrenia. Commonly used measures of psychopathology used in law enforcement research include the Minnesota Multiphasic Personality Inventory (MMPI; MMPI-2), Inwald Personality Inventory (IPI), Millon Multiaxial Clinical Inventory (MMCI-III), Personality Assessment Inventory (PAI), and the Clinical Assessment Questionnaire (CAQ). Such measures are designed to "screen out" applicants who have psychological problems that would cause performance or discipline problems on the job. Measures of psychopathology are not designed to "select in" applicants, and as will be seen in the meta-analysis results, they are seldom predictive of job performance.

Measures of psychopathology are considered to be medical exams. As a result, to be in compliance with the Americans with Disabilities Act (ADA), they can only be administered after a conditional offer of employment has been made. Prior to the passage of the ADA in 1990, these tests were routinely administered to all applicants. As a result, the type of data available after 1990 is different from that available prior to that date. Prior to 1990, data on entire applicant pools were available whereas after 1990, the only psychopathology test data available are scores from current officers or the few applicants who were offered jobs. As a result, one would expect scores from post-1990 studies to be relatively normal.

Minnesota Multiphasic Personality Inventory (MMPI)

The MMPI is the most commonly used and researched measure of psychopathology in law enforcement selection. The 556-item MMPI was first published in 1943 and the 567-item revised version (MMPI-2) in 1989. Test takers indicate whether each of the items is true or false (e.g., "Most of the time I wish I were dead"). The MMPI yields scores on three validity scales (L, F, K) designed to evaluate how honestly and reliably a person responded to the questions, 10 primary clinical scales (Hs, D, Hy, Pd, Mf, Pa, Pt, Sc, Ma, Si), and hundreds of research and special scales. The MMPI-2 yields scores on the same scales as well as three additional validity scales (Fb, VRIN, TRIN). Descriptions of the scales are located in Appendix A at the end of this chapter.

Scores on the MMPI and MMPI-2 are reported as T scores that have a mean of 50 and a standard deviation of 10. Scores above 70 on the MMPI and 65 on the MMPI-2 are thought to be clinically elevated. The reliability of scores from the MMPI and MMPI-2 is typical for personality inventories with a median test-retest coefficient of .74 for the MMPI and .81 for the MMPI-2 (Graham, 1993). Median internal consistency coefficients are .87 for the MMPI and .64 for the MMPI-2 (Graham, 1993).

Inwald Personality Inventory (IPI)

The Inwald Personality Inventory (IPI) uses 310 true-false items to measure a variety of behavioral and emotional problems. In addition to a scale that that taps defensiveness in taking the inventory (Guardedness; GD), the 25 constructs measured by the IPI are:

Acting out behaviors: Alcohol use (AL), Drug use (DG), Driving violation (DV), Job difficulties (JD), Trouble with society and the law (TL), and Absence Abuse (AA)

Acting out attitudes: Substance abuse (SA), Antisocial attitudes (AS), Hyperactivity (HP), Rigid type (RT), and Type A (TA)

Internalized conflict: Illness concerns (IC), Treatment programs (TP), Anxiety (AN), Phobic personality (PH), Obsessive personality (OB), Depression (DE), Loner type (LO), Unusual experience and thoughts (EU)

Interpersonal conflict: Lack of assertiveness (LA), Interpersonal difficulties (ID), Undue suspiciousness (US), Family conflicts (FC), Sexual concerns (SC), and Spouse/mate conflicts (SP)

As with the MMPI, the IPI scores are reported as *T* scores (Mean of 50, standard deviation of 10).

Millon Clinical Multiaxial Inventory (MCMI)

The Millon Clinical Multiaxial Inventory (MCMI) is a 175-item inventory designed to measure aspects of psychopathology. Though the MCMI was designed for use with psychiatric patients, it has been used with police applicants in several studies. Since its inception, there have been two revisions of the inventory and the three versions are referred to as the MCMI-I, MCMI-II, and MCMI-III respectively.

Unlike the MMPI, the MCMI uses base rate scores (BR) rather than *T* scores. A BR score of 35 represents the median for the normal population, 60 the median for psychiatric patients, 75 as the point at which a particular disorder can be said to be present, and 85 as the point at which the disorder is the most dominant for the individual.

The MCMI yields scores on:

- 4 validity/response set scales (validity, disclosure, desirability, debasement)

- 8 personality type scales (schizoid, avoidant, dependent, histrionic, narcissistic, antisocial, compulsive, negativistic)
- 6 severe personality scales (depressive, aggressive, self-defeating, schizotypal, borderline, paranoid)
- 6 clinical syndrome scales (anxiety disorder, somatoform disorder, hypomania, dysthymic disorder, alcohol dependence, drug dependence, post traumatic stress)
- 3 severe clinical syndrome scales (thought disorders, major depression, delusional disorders).

Clinical Analysis Questionnaire (CAQ)

The Clinical Analysis Questionnaire (CAQ) contains a measure of normal personality (short form of the 16PF) as well as 144 items tapping the following 12 clinical dimensions: Hypochondriasis, suicidal depression, agitation, anxious depression, low energy depression, guilt and resentment, social introversion, paranoia, psychopathic deviate, schizophrenia, psychasthenia, and psychological inadequacy.

CAQ scores are reported as STEN scores: a standard score with a mean of 5.5 and a standard deviation of 2. The temporal stability of CAQ scores ranges from .67 to .90.

Personality Assessment Inventory (PAI)

The Personality Assessment Inventory (PAI) is a 344-item inventory designed to identify psychopathology. The PAI can be completed in about 50 minutes and contains the following scales:

Validity scales: Inconsistency, infrequency, negative impression, and positive impression

Clinical scales: Somatic complaints (SOM), anxiety (ANX), anxiety related disorders (ARD), depression (DEP), mania (MAN), paranoia (PAR), schizophrenia (SCZ), borderline personality (BOR), antisocial personality (ANT), aggression (AGG), and substance abuse (DRG)

Treatment consideration scales: Aggression (AGG), suicidal ideation (SUI), stress (STR), non-support (NON), and treatment rejection (RXR)

Interpersonal scales: Dominance (DOM) and warmth (WRM)

Scores on the PAI have demonstrated acceptable reliability with a median coefficient alpha of .81 and a median test-retest reliability of .83.

Measures of Normal Personality

Tests of normal personality measure the traits exhibited by normal individuals in everyday life. Examples of such traits are extraversion, shyness, assertiveness, and friendliness. Though there is some disagreement, psychologists today generally agree there are five main personality dimensions. Popularly known as the *Big Five*, these dimensions are openness to experience (bright, adaptable, inquisitive), conscientiousness (reliable, dependable, rule oriented), extraversion (outgoing, friendly, talkative), agreeableness (works well with others, loyal), and emotional stability (calm, not anxious or tense). Commonly used measures of normal personality in law enforcement research include the California Psychological Inventory (CPI), 16PF, and Edwards Personal Preference Schedule (EPPS). In contrast to measures of psychopathology, measures of normal personality are used to "select in" rather than "screen out" applicants.

California Psychological Inventory (CPI)

The CPI is one of the oldest and most respected personality inventories. The current version (Form 434) has 434 items and is 28 items shorter than the previous version (items were eliminated to avoid problems with the Americans with Disabilities Act and other employment laws) and 44 items shorter than the original version.

Answers to the 434 items yield scores on 20 dimensions in four primary areas:

Measures of poise: dominance, capacity for status, sociability, social presence, self-acceptance, independence, and empathy.

Measures of normative orientation and values: Responsibility, socialization, self-control, good impression, communality, well-being, and tolerance.

Measures of cognitive and intellectual functioning: Achievement via conformance, achievement via independence, and intellectual efficiency.

Measures of role and interpersonal style: Psychological mindedness, flexibility, and femininity.

The CPI also produces scores on several special scales including managerial potential (Mp), work orientation (Wo), creative temperament (Ct), leadership potential (Lp), amicability (Ami), law enforcement orientation (Leo), anxiety (Anx), narcissism (Nar), and tough-mindedness (Tm).

Similar to the MMPI, scores from the CPI are reported as T scores: a standard score with a mean of 50 and a standard deviation of 10. Coefficient alphas for the scales range from .62 (psych mindedness) to .84 (well-being) with a median of .77. Test-retest coefficients range from .51 (communality) to .84 (femininity) with a median of .68.

16PF

The 16PF is a 185-item personality inventory that can be completed in 25-50 minutes. The 16PF was first published in 1949 and its 4th (1968) and 5th (1993) versions were the ones used in the law enforcement research cited in this project. The inventory has three scales to help interpret the validity of the responses to the items: Impression management (IM), infrequency (INF), and acquiescence (ACQ). If these three scales suggest an invalid response style, the other scales are not interpreted.

The names of the 16 primary factors have changed over the years and are usually reported as opposites (e.g., warm vs. reserved). High scores indicate a tendency toward the first word in the pair (i.e., warm). The most recent primary-factor names (with previous names in parentheses) are:

- A: Warm (outgoing) vs. reserved
- B: Abstract reasoning (bright, intelligent) vs. concrete-reasoning
- C: Emotional stability (calm, stable) vs. reactive
- E: Dominant (assertive) vs. deferential
- F: Lively (happy-go-lucky, enthusiastic, impulsive, surgent) vs. serious
- G: Rule-conscious (conscientious, conformity) vs. expedient
- H: Socially bold (venturesome, bold) vs. shy
- I: Sensitive (tender-minded) vs. utilitarian
- L: Vigilant (suspicious, skeptical) vs. trusting
- M: Abstract (imaginative) vs. grounded
- N: Private (shrewd, closed-mouthed) vs. forthright
- O: Apprehensive (insecure, worrisome) vs. self-assured
- Q1: Open to Change (experimenting, radical) vs. traditional
- Q2: Self-Reliant (self-sufficient, self-directed) vs. group-oriented
- Q3: Perfectionistic (controlled, disciplined) vs. tolerates disorder
- Q4: Tense

These 16 primary factors fall within 5 global factors:

- Extraversion
- Anxiety
- Tough-mindedness (tough poise)
- Independence
- Self-control (control)

16-PF scores are reported as STEN scores: a standard score with a mean of 5.5 and a standard deviation of 2. Test-retest reliability coefficients for the five global factor scores range from

.84 to .91 and from .69 to .87 for the 16 primary factors. The median internal-consistency coefficient (alpha) was .74.

Edwards Personal Preference Schedule (EPPS)

The EPPS was developed in 1959 on the basis of the personality theory of H. S. Murray. The EPPS consists of 225 pairs of psychological needs and yields a test consistency score as well as scores on 15 dimensions of manifest needs: achievement, deference, order, exhibition, autonomy, affiliation, intraception, succorance, dominance, abasement, nurturance, change, endurance, heterosexuality, and aggression. Law enforcement research using the EPPS has been fairly limited. Azen, Snibbe, Montgomery, Fabricatore, and Earle (1974) found a correlation of -.27 between the intraception scale and tenure, Balch (1977) found that higher scores on need for achievement (r = .37) and consistency in completing the test (r = .22) were associated with completing a police academy, and Shaffer (1996) found that higher scores on the dominance scale were related to increased traffic accidents and decreased damage claims. Sheppard, Bates, Fracchia, and Merlis (1974) and Simon, Wilde, and Cristal (1973) compared EPPS means for police officers with the means for adult males and college males.

Employee Personality Inventory (EPI)

Though not designed for employee selection purposes, the Employee Personality Inventory was used in three studies cited in this book: Raynes (1997), Schelling (1993), and Surrette Ebert, Willis, & Smallidge (2003). The EPI contains 40 pairs of traits and applicants are asked to select which of the two traits is most like them. The EPI yields scores on five dimensions: Thinking (curious, creative, smart), Directing (aggressive, assertive, dominant), Communicating (outgoing, friendly, talkative), Soothing (agreeable, calm, helpful), and Organizing (organized, neat, careful). The median test-retest reliability for scores on the five scales is .85.

Guildford-Zimmerman Temperament Survey

The Guilford-Zimmerman Temperament Survey (GZTS) yields scores on 10 dimensions: general activity, restraint, ascendance, sociability, emotional stability, objectivity, friendliness, thoughtfulness, personal relations, and masculinity. The 300 items in the survey can be completed in about 45 minutes. Internal consistency coefficients are in the .80s but the test-retest reliability is fairly low for this inventory.

META-ANALYSIS RESULTS

Meta-analyses were conducted separately for each of the commonly used inventories for which there were sufficient data (MMPI, IPI, CPI, 16PF), separately for types of psychopathology (e.g., depression, hysteria), and separately for the Big 5 categories of normal personality (openness to experience, conscientiousness, extraversion, agreeableness, and emotional stability). For the meta-analyses involving a particular scale from a single inventory (e.g., the L scale from the MMPI), no corrections were made for predictor unreliability because the purpose of these meta-analyses was to establish the validity of a particular scale rather than a construct measured by a variety of scales with differing reliabilities. At the end of the chapter, a "police profile" of average scores on the major personality inventories is included.

Commonly Used Inventories: Individual Scale Scores

This section contains the meta-analysis results of the validity of the CPI, IPI, MMPI/MMPI-2, and 16PF. There were an insufficient number of studies to conduct meta-analyses on the other personality inventories. Conducting a meta-analysis of the individual scale scores was somewhat problematic in that many studies did not provide a complete set of correlations. Instead, a common practice was to list only those correlations that were statistically significant. Such a practice, combined with the small number of studies in some of the analyses, dictates caution in interpreting the meta-analysis tables as some of the coefficients are probably upwardly biased. To reduce this bias, attempts were made to obtain complete sets of data. Hilson Research (the publisher of the IPI) was contacted twice as were authors of several of the studies. Only two researchers responded to the request for the complete set of correlations.

MMPI and MMPI-2

As shown in Table 7.1 and 7.2, all of the correlations involving individual MMPI scales and measures of academy and patrol performance are low, and the great majority of the correlations are not statistically significant. As shown in Tables 7.3 and 7.4, no individual scales significantly predict discipline problems or commendations. Because the F scale is comprised of items from the other clinical scales and because it is significantly related to both academy performance (r = -.11) and supervisor ratings of performance (r = -.09), it is probably the most useful individual MMPI scale.

Table 7.1: Meta-analysis results for the validity of the MMPI in predicting academy grades

MMPI Scale	K	N	r	95% Confidence Interval		ρ	90% Credibility Interval		Var	Q_w
				Lower	Upper		Lower	Upper		
L	9	1,469	-.02	-.11	.07	-.03	-.32	.25	30%	29.57*
F	9	1,469	-.11	-.17	-.04	-.16	-.31	.00	61%	14.82
K	8	1,364	.08	.02	.14	.12	-.01	.24	71%	11.31
Hs	6	973	-.09	-.15	-.02	-.13	-.13	-.13	100%	1.00
D	7	1,073	-.07	-.13	-.01	-.11	-.11	-.11	100%	2.80
Hy	7	1,073	.02	-.04	.08	.04	.04	.04	100%	4.66
Pd	7	1,105	-.04	-.10	.02	-.06	-.06	-.06	100%	1.65
MF	9	1,411	-.02	-.10	.05	-.04	-.27	.20	42%	21.66*
Pa	8	1,387	.04	-.01	.09	.06	-.05	.16	77%	10.38
Pt	7	1,105	-.03	-.09	.03	-.05	-.05	-.05	100%	3.46
Sc	6	973	-.07	-.14	-.01	-.11	-.11	-.11	100%	5.04
Ma	6	973	-.11	-.20	-.02	-.16	-.40	.08	40%	15.01*
Si	9	1,478	-.01	-.11	.09	-.02	-.36	.32	24%	37.44*

K=number of studies, N=sample size, r = mean correlation, ρ = mean correlation corrected for range restriction, criterion unreliability, and predictor reliability, VAR = percentage of variance explained by sampling error and study artifacts, Q_w = the within group heterogeneity

Table 7.2: Meta-analysis results for the validity of the MMPI in predicting supervisor ratings of performance

MMPI Scale	K	N	*r*	95% Confidence Interval		*ρ*	90% Credibility Interval		Var	Q_w
				Lower	Upper		Lower	Upper		
L	25	3.279	- .03	- .07	.00	- .05	- .08	- .03	49%	51.14*
F	23	3,304	- .09	- .12	- .05	- .15	- .39	.10	49%	46.99*
K	26	3,519	.04	- .04	.11	.06	- .24	.36	39%	67.03*
Hs	24	2,663	- .02	- .09	.04	- .04	- .27	.20	56%	41.14*
D	23	2,715	- .06	- .11	- .01	- .10	- .24	.04	77%	29.89*
Hy	24	3,222	.02	- .04	.08	.03	- .19	.26	54%	44.60*
Pd	24	3,273	- .08	- .15	- .01	- .14	- .16	- .11	49%	49.16*
MF	21	2,768	- .06	- .10	- .03	- .11	- .20	- .02	89%	23.71
Pa	27	3,314	- .01	- .08	.07	- .01	- .29	.27	45%	60.56*
Pt	22	2,585	- .07	- .13	- .01	- .12	- .31	.07	63%	33.68*
Sc	22	2,585	- .09	- .17	- .01	- .15	- .45	.16	43%	50.81*
Ma	24	3,204	- .09	- .16	- .03	- .16	- .40	.08	52%	45.84*
Si	23	2,861	- .01	- .05	.02	- .02	- .16	.12	76%	30.39

K=number of studies, N=sample size, r = mean correlation, ρ = mean correlation corrected for range restriction, criterion unreliability, and predictor reliability, VAR = percentage of variance explained by sampling error and study artifacts, Q_w = the within group heterogeneity

Table 7.3: Meta-analysis results for the validity of the MMPI in predicting discipline problems and complaints

MMPI Scale	K	N	r	95% Confidence Interval		ρ	90% Credibility Interval		Var	Q_w
				Lower	Upper		Lower	Upper		
L	11	4,967	- .02	- .07	.02	- .04	- .21	.13	35%	31.63*
F	10	3,620	.01	- .03	.06	.02	- .10	.14	58%	17.14*
K	11	3,695	.00	- .03	.03	.00	.00	.00	100%	7.97
Hs	13	3,814	- .02	- .06	.02	- .03	- .12	.07	72%	18.06
D	12	3,712	- .01	- .05	.03	- .02	- .13	.08	68%	17.77
Hy	12	3,976	.00	- .05	.05	- .01	- .13	.12	59%	20.51*
Pd	14	4,143	.03	- .02	.08	.05	- .14	.24	41%	34.56*
MF	11	3,647	.00	- .04	.04	.00	- .10	.10	67%	16.38
Pa	11	3,647	.01	- .03	.06	.02	- .12	.17	52%	21.13*
Pt	13	3,814	- .02	- .05	.01	- .03	- .03	- .03	100%	11.52
Sc	13	3,797	.00	- .03	.03	.01	- .05	.06	90%	14.51
Ma	12	3,749	.02	- .02	.05	.03	.03	.03	100%	11.56
Si	13	3,813	.01	- .03	.04	.01	.01	.01	100%	11.97

K=number of studies, N=sample size, r = mean correlation, ρ = mean correlation corrected for range restriction, criterion unreliability, and predictor reliability, VAR = percentage of variance explained by sampling error and study artifacts, Q_w = the within group heterogeneity

Table 7.4: Meta-analysis results for the validity of the MMPI in predicting citizen and department commendations

MMPI Scale	K	N	r	95% Confidence Interval		ρ	90% Credibility Interval		Var	Q_w
				Lower	Upper		Lower	Upper		
L	6	727	- .01	- .09	.06	- .02	- .02	- .02	100%	1.93
F	6	727	- .01	- .10	.08	- .02	- .22	.19	50%	12.03*
K	6	727	- .04	- .12	.04	- .05	- .18	.08	71%	8.40
Hs	7	754	- .05	- .16	.06	- .07	- .40	.20	33%	21.50*
D	7	754	.02	- .05	.09	.03	.03	.03	100%	6.28
Hy	8	1,083	- .01	- .09	.08	- .01	- .06	.04	43%	18.78*
Pd	7	754	- .06	- .18	.07	- .08	- .42	.27	28%	25.29*
MF	7	754	- .02	- .11	.07	- .02	- .21	.16	57%	12.36*
Pa	7	754	- .01	- .10	.08	- .01	- .22	.19	52%	13.40*
Pt	7	754	- .07	- .14	.00	- .10	- .39	.19	36%	19.47*
Sc	7	754	- .05	- .16	.07	- .06	- .37	.25	33%	21.50*
Ma	7	754	- .01	- .08	.06	- .02	- .13	.09	79%	8.81
Si	7	754	- .03	- .11	.06	- .04	- .20	.13	64%	10.90

K=number of studies, N=sample size, r = mean correlation, ρ = mean correlation corrected for range restriction, criterion unreliability, and predictor reliability, VAR = percentage of variance explained by sampling error and study artifacts, Q_w = the within group heterogeneity

Table 7.5: Meta-analysis results for the validity of the MMPI in predicting absenteeism

MMPI Scale	K	N	r	95% Confidence Interval		ρ	90% Credibility Interval		Var	Q_w
				Lower	Upper		Lower	Upper		
L	5	1,439	.03	-.02	.07	.04	.04	.04	100%	1.75
F	6	1,768	.01	-.04	.05	.01	.01	.01	100%	2.69
K	5	1,439	-.05	-.10	.00	-.07	-.09	-.05	98%	5.06
Hs	6	1,529	-.05	-.10	.00	-.07	-.13	.00	81%	7.45
D	5	1,439	-.03	-.09	.02	-.05	-.05	-.05	100%	1.72
Hy	5	1,439	-.06	-.11	-.01	-.08	-.08	-.08	100%	1.31
Pd	6	1,768	.00	-.05	.04	.00	-.08	.07	76%	7.85
MF	6	1,768	.00	-.04	.05	.01	.02	.01	100%	5.75
Pa	5	1,439	-.03	-.08	.02	-.04	-.04	-.04	100%	3.92
Pt	5	1,439	-.05	-.10	.01	-.06	-.06	-.06	100%	4.88
Sc	5	1,439	-.06	-.11	-.01	-.08	-.08	-.08	100%	2.18
Ma	5	1,439	.01	-.04	.06	.00	.00	.03	98%	5.08
Si	6	1,530	-.01	-.06	.04	-.02	-.02	-.02	100%	1.86

K=number of studies, N=sample size, r = mean correlation, ρ = mean correlation corrected for range restriction, criterion unreliability, and predictor reliability, VAR = percentage of variance explained by sampling error and study artifacts, Q_w = the within group heterogeneity

Inwald Personality Inventory (IPI)

The meta-analysis results shown in Table 7.6 suggest that several of the IPI scales may be useful predictors of police performance. However, most of the predictors with the highest correlations (i.e., job difficulties, trouble with the law, and absence abuse) represent information that would already be obtained during a background check. Thus, by the time the IPI would be administered after a conditional offer of employment, applicants with previous work and legal problems would have already been screened out.

California Psychological Inventory (CPI)

As shown in Tables 7.7, 7.8, and 7.9, there are several CPI scales that are significantly related to supervisor ratings of performance, academy performance, or discipline problems. The scales that stand out are tolerance and intellectual efficiency as both are significantly correlated to academy grades, supervisor ratings, and discipline problems. People scoring high in tolerance are tolerant, non-judgmental, and resourceful and those scoring high in intellectual efficiency are intelligent, clear thinking, and capable.

16PF

Seven studies investigated the validity of at least one scale of the 16PF. As can be seen in Table 7.10, most of the scales were investigated in only 3 studies. Of the 16 scales, only two—dominance and conscientiousness—had uncorrected correlations that were significantly related to performance. People scoring high in dominance are assertive, unconventional, and competitive and those scoring high in conscientiousness are conscientious, persistent, and responsible.

Table 7.6: Meta-analysis results for the validity of the IPI in predicting performance ratings

IPI Scale	K	N	r	95% Confidence Interval		ρ	90% Credibility Interval		Var	Q_w
				Lower	Upper		Lower	Upper		
Guardedness	6	1,361	-.02	-.09	.04	-.04	-.19	.10	62%	9.73
Alcohol Use	6	908	-.02	-.12	.07	-.04	-.27	.19	49%	12.38*
Drug Use	5	765	-.03	-.11	.04	-.06	-.06	-.06	100%	1.87
Driving Violations	4	668	-.07	-.15	.00	-.13	-.13	-.13	100%	3.50
Job Difficulties	5	765	-.19	-.26	-.12	-.32	-.32	-.32	100%	3.43
Trouble with the Law	6	1,361	-.16	-.21	-.11	-.27	-.27	-.27	100%	3.77
Absence Abuse	7	1,504	-.13	-.21	-.06	-.23	-.40	-.05	60%	11.67
Substance Abuse	5	765	-.14	-.22	-.05	-.23	-.34	-.12	83%	6.00
Antisocial Attitudes	5	765	-.19	-.30	-.07	-.31	-.56	-.07	53%	9.45
Hyperactivity	5	765	-.16	-.25	-.06	-.26	-.41	-.11	74%	6.79
Rigid Type	7	1,504	-.07	-.12	-.02	-.13	-.21	-.05	86%	8.18
Type A	4	668	-.09	-.16	-.01	-.15	-.15	-.15	100%	2.01
Illness Concerns	5	1,264	-.10	-.17	-.03	-.17	-.28	-.05	72%	6.97

K=number of studies, N=sample size, r = mean correlation, ρ = mean correlation corrected for range restriction, criterion unreliability, and predictor reliability, VAR = percentage of variance explained by sampling error and study artifacts, Q_w = the within group heterogeneity

Table 7.6: Meta-analysis results for the validity of the IPI in predicting performance ratings (continued)

IPI Scale	K	N	r	95% Confidence Interval		ρ	90% Credibility Interval		Var	Q_w
				Lower	Upper		Lower	Upper		
Treatment Programs	5	1,264	-.01	-.06	.05	-.01	-.01	-.01	100%	4.66
Anxiety	5	811	-.09	-.16	-.02	-.15	-.15	-.15	100%	0.61
Phobic Personality	5	811	-.05	-.14	.05	-.08	-.29	.12	54%	9.28
Obsessive Personality	5	1,264	-.10	-.16	-.05	-.17	-.17	-.17	100%	2.64
Depression	4	668	-.16	-.27	-.04	-.26	-.48	-.04	54%	7.42
Loner	6	908	-.15	-.25	-.05	-.25	-.50	.00	50%	12.01*
Unusual Experiences	5	765	-.14	-.31	.03	-.23	-.71	.25	21%	24.10*
Lack of Assertiveness	6	908	.00	-.08	.09	.00	-.18	.18	61%	9.83
Interpersonal Difficulty	4	668	-.19	-.31	-.07	-.32	-.53	-.12	61%	6.61
Undue Suspiciousness	6	1,361	-.15	-.24	-.05	-.25	-.49	.01	42%	14.17*
Family Concerns	5	765	-.13	-.24	-.02	-.22	-.46	.02	50%	9.91*
Sexual Concerns	4	668	-.13	-.21	-.06	-.22	-.22	-.22	100%	3.74
Spouse Conflicts	4	668	-.13	-.21	-.06	-.22	-.22	-.22	100%	2.18

K=number of studies, N=sample size, r = mean correlation, ρ = mean correlation corrected for range restriction, criterion unreliability, and predictor reliability, VAR = percentage of variance explained by sampling error and study artifacts, Q_w = the within group heterogeneity

Table 7.7: Meta-analysis results for the validity of the CPI in predicting performance ratings

CPI Scale	K	N	r	95% Confidence Interval		ρ	90% Credibility Interval		Var	Q_w
				Lower	Upper		Lower	Upper		
Dominance	14	1,117	.05	- .01	.11	.08	- .07	.23	82%	17.05
Capacity for Status	13	1,072	.06	.00	.12	.09	.09	.09	100%	10.41
Sociability	14	1,117	.03	- .03	.09	.05	- .06	.17	87%	16.01
Social Presence	13	1,072	.06	.08	.12	.10	.10	.10	100%	5.92
Self-acceptance	15	1,166	.01	- .05	07	.02	- .29	.33	51%	29.63*
Well-being	16	1,256	.15	.10	.21	.26	.26	.26	100%	14.83
Responsibility	17	1,400	.12	.07	.17	.21	.07	.34	84%	20.12
Socialization	14	1,121	.10	.03	.17	.16	.00	.33	79%	17.80
Self-control	15	1,187	.16	.08	.24	.26	.00	.53	61%	24.48*
Tolerance	15	1,187	.20	.15	.26	.34	.34	.34	100%	9.67
Good Impression	13	1,072	.10	.01	.19	.17	- .15	.48	51%	25.62*
Communality	16	1,186	.11	.05	.16	.18	.07	.29	90%	17.74
Ach via Conformity	15	1,261	.17	.11	.22	.28	.00	.54	58%	26.01*
Ach via Independence	15	1,261	.12	.05	.20	.21	- .01	.43	66%	22.68
Intellectual Efficiency	13	1,072	.14	.04	.23	.23	- .13	.59	44%	29.78*
Psych Mindedness	13	1,072	.12	.06	.18	.21	.21	.21	100%	11.56
Flexibility	14	1,102	.05	- .01	.11	.08	.08	.08	100%	13.40
Femininity	14	1,102	.09	.04	.15	.16	.16	.16	100%	13.01
V1	3	277	.11	- .01	.22	.18	.18	.18	100%	1.89
V2	3	277	.12	- .06	.29	.20	- .11	.50	50%	6.04
V3	3	277	.07	- .05	.19	.12	.12	.12	100%	1.67

Table 7.8: Meta-analysis results for the validity of the CPI in predicting academy performance

CPI Scale	K	N	r	95% Confidence Interval		ρ	90% Credibility Interval		Var	Q_w
				Lower	Upper		Lower	Upper		
Dominance	7	732	.20	.13	.27	.30	.30	.30	100%	1.19
Capacity for Status	8	777	.23	.16	.30	.33	.33	.33	100%	5.75
Sociability	8	777	.24	.17	.30	.34	.34	.34	100%	4.89
Social Presence	6	626	.18	.11	.26	.26	.26	.26	100%	4.57
Self-acceptance	4	488	.20	.12	.29	.29	.29	.29	100%	3.80
Well-being	3	386	.26	.16	.35	.37	.31	.43	95%	3.17
Responsibility	2	343	.20	.09	.30	.29	.29	.29	100%	0.23
Socialization	2	343	.06	- .05	.17	.09	.09	.09	100%	0.80
Self-control	2	343	.02	- .09	.13	.03	.03	.03	100%	0.00
Tolerance	4	554	.26	.19	.34	.38	.14	.62	50%	8.08*
Good Impression	2	343	.02	- .09	.12	.02	.02	.02	100%	0.01
Communality	4	491	.12	.03	.21	.18	.18	.18	100%	3.62
Ach via Conformity	4	554	.19	.11	.27	.27	.27	.27	100%	3.15
Ach via Independence	4	554	.27	.12	.42	.38	.11	.66	42%	9.47*
Intellectual Efficiency	7	732	.32	.23	.40	.45	.45	.45	100%	6.92
Psych Mindedness	4	554	.16	.05	.28	.24	.06	.41	61%	6.57
Flexibility	2	343	.17	- .06	.34	.21	.21	.21	31%	6.42*
Femininity	2	343	- .18	- .28	- .08	- .27	- .09	.50	100%	0.32

Table 7.9: Meta-analysis results for the validity of the CPI in predicting discipline problems

CPI Scale	K	N	r	95% Confidence Interval		ρ	90% Credibility Interval		Var	Q_w
				Lower	Upper		Lower	Upper		
Dominance	6	939	-.07	-.20	.05	-.12	-.47	.23	28%	21.37*
Capacity for Status	7	1,041	.03	-.06	.11	.04	-.19	.27	48%	14.66*
Sociability	5	849	.01	-.06	.07	.01	.01	.01	100%	0.86
Social Presence	5	849	.01	-.06	.08	.01	.01	.01	100%	2.95
Self-acceptance	5	849	.03	-.04	.10	.05	.05	.05	100%	2.31
Well-being	7	1,530	-.12	-.20	-.05	-.20	-.38	-.02	54%	12.89*
Responsibility	8	1,620	-.09	-.17	.00	-.14	-.41	.13	34%	23.46*
Socialization	6	1,428	-.18	-.26	-.11	-.29	-.43	-.16	70%	8.61
Self-control	8	1,620	-.17	-.26	-.07	-.27	-.57	.03	33%	24.51*
Tolerance	7	1,428	-.15	-.20	-.10	-.24	-.24	-.24	100%	5.50
Good Impression	8	1,620	-.07	-.16	.01	-.12	-.39	.15	33%	24.06*
Communality	8	1,620	-.12	-.17	-.07	-.20	-.20	-.20	100%	4.57
Ach via Conformity	6	951	-.04	-.10	.03	-.06	-.06	-.06	100%	5.10
Ach via Independence	5	849	-.11	-.18	-.05	-.19	-.25	-.12	92%	5.45
Intellectual Efficiency	5	849	-.04	-.10	.03	-.06	-.16	.05	80%	6.23
Psych Mindedness	5	849	-.06	-.15	.03	-.10	-.27	.07	61%	8.19
Flexibility	5	849	-.09	-.19	.01	-.15	-.38	.09	46%	10.85*
Femininity	5	849	-.05	-.12	.02	-.08	-.08	-.08	100%	3.92

Table 7.10: Meta-analysis results for the validity of the 16PF in predicting supervisor ratings of performance

16PF Scale	K	N	r	95% Confidence Interval		ρ	90% Credibility Interval		Var	Q_w
				Lower	Upper		Lower	Upper		
Outgoing	3	298	-.04	-.15	.07	-.07	-.07	-.07	100%	2.56
Bright	3	298	.14	-.02	.29	.23	-.17	.62	35%	8.45*
Calm	3	298	.06	-.10	.22	.10	-.13	.32	62%	4.87
Dominant	6	1,016	.13	.07	.19	.23	.14	.31	89%	6.75
Happy	3	298	.08	-.03	.19	.14	.14	.14	100%	1.97
Conscientious	7	1,127	.15	.07	.23	.25	.09	.41	70%	10.01
Venturesome	5	683	.01	-.06	.09	.02	-.04	.08	93%	5.35
Tender-minded	3	298	-.03	-.14	.08	-.05	-.05	-.05	100%	1.96
Suspicious	3	298	-.06	-.17	.06	-.10	-.10	-.10	100%	0.19
Imaginative	3	298	-.05	-.21	.11	-.08	-.49	.02	32%	9.25*
Shrewd	3	298	-.01	-.12	.11	-.02	-.02	-.02	100%	2.97
Apprehensive	3	298	-.11	-.22	.00	-.18	-.18	-.18	100%	0.84
Q!: Experimenting	4	434	-.08	-.18	.01	-.14	-.14	-.14	100%	3.91
Q2: Self-directed	3	298	.07	-.04	.19	.12	.12	.12	100%	2.60
Q3: Disciplined	5	683	.03	-.08	.13	.05	-.16	.25	57%	8.79
Q4: Tense	5	683	.02	-.06	.09	.03	.03	.03	100%	3.35

MMPI and MMPI-2 Patterns

It is probably not surprising that individual scales of the MMPI had such low correlations with measures of police performance because the MMPI scales are usually interpreted with cutoff scores rather than with linear relationships. So, what these results basically mean is that score differences within the normal range of a single scale (30-69 for the MMPI and 35-64 for the MMPI-2) are not particularly useful. Because applicants with extremely high scores (above 70 for the MMPI, above 65 for the MMPI-2) are seldom hired, it is impossible from the available data to determine how these officers would perform. There were not enough studies using the MMPI-2 to look at possible validity differences between the two versions of the test.

As mentioned previously, using MMPI scores within a normal range is not a common or useful practice. Instead, psychologists look at extreme scores or patterns of scores. In the literature, there were four methods used to interpret MMPI patterns: Good Cop/Bad Cop Profile, Goldberg Index, Husemann Index, and the Gonder Index. Because there were few studies investigating the various scale configurations, data from Surrette, Aamodt, and Serafino (1990) were reanalyzed to increase the number of studies. Robert Davis and Cary Rostow from Matrix, Inc., a Louisiana-based company specializing in law enforcement screening, were kind enough to lend me their extensive data set so that further analyses could be conducted on these scale configurations.

Good Cop/Bad Cop Profile

The Good Cop/Bad Cop Profile was developed by Blau, Super, and Brady (1993). A prediction of an applicant being a "good cop" is made when the applicant's T scores are less than 60 on the Hy, Hs, Pd, and Ma scales and less than 70 on the other clinical scales. As shown in Table 7.11, evidence of validity for this scale across the four studies is promising but further research is needed.

Goldberg Index

The formula for the Goldberg Index is L+Pa+Sc-Hy-Pt. Only three studies investigated the validity of this combination of scales. Costello, Schneider, Schoenfeld, and Kobos (1982) found a correlation of - .28 between Goldberg scores and performance, a reanalysis of the Surrette, Aamodt, and Serafino (1990) data yielded a correlation of - .04 with performance ratings, and the Matrix data yielded a correlation of .14 with being terminated for cause.

Husemann Index

The Husemann Index is a measure of aggression and is formed by summing the F, Pd, and Ma scales. Costello and Schneider (1996) used a cutoff score of 192 and found a correlation of .22 between index category (above 192, less than 192) and being categorized as a problem officer (officers in the top 10% of days suspended) or a non-problem officer (officer in the bottom 10% of days suspended). Surrette et al. (1990) found a correlation of - .10 between Husemann Index scores and ratings of patrol performance. The Matrix data revealed a significant correlation (r = .15) between Husemann scores and being terminated for cause. As with the Good Cop/Bad Cop profile, the Husemann Index appears to have promise but needs further research.

Gonder Index

The Gonder Index is created by summing the Pd, Pt, Mf, Ma, Hs, and Hy scales. In the only study looking at this combination, Gonder (1998) found a correlation of .02 between the index scores and completion of the academy (cadets completing the academy had slightly higher scores). A reanalysis of the Surrette et al. (1990) data yielded a correlation of .04 between the Gonder Index and ratings of patrol performance. On the basis of these two studies, this combination of scales does not appear promising.

Table 7.11

Validity of MMPI profile configurations

Method/ Study	Criterion	N	Base Rate	Overall Prediction Accuracy %	r
Good Cop/Bad Cop					
Blau et al. (1993)	Performance Ratings	30	50.0	80.0	- .76
Brewster & Stoloff (1999)	Performance Ratings	39	79.0	82.1	- .44
Surrette et al. (1990)	Performance Ratings	129			- .03
Matrix data	Termination for Cause	1,970	92.3	88.7	.15
Goldberg Index					
Costello et al. (1982)	Performance Ratings	161	85.7	80.7	- .28
Surrette et al. (1990)	Performance Ratings	129			- .04
Matrix data	Termination for Cause	1,970			.14
Husemann Index					
Costello (1996)	Suspensions	107	89.7	91.0	.22
Surrette et al. (1990)	Performance Ratings	129			- .10
Matrix data	Termination for Cause	1,970			.20
Gonder Index					
Surrette et al. (1990)	Performance Ratings	129			.04
Gonder (1998)	Academy Grades	291			.02

Clinical Interpretation of Personality Inventories

Several studies investigated the validity of a psychologist's interpretation of the MMPI/MMPI-2, CPI, or IPI. Unfortunately, we do not know the criteria used by these psychologists to determine if an applicant was "psychologically unfit." That is, did they use a combination of the scales such as those previously discussed or did the psychologist use a rule such as disqualifying any applicant with a clinically elevated score on any scale? A further confound in determining the validity of clinical interpretation of personality inventories is that the interpretations were often made after reviewing other materials such as cognitive ability scores and background questionnaires.

Thus, in many cases, it is impossible to determine if a large validity coefficient can be attributed to the interpretation of the personality inventory, a cognitive ability test, or some combination.

Regression of Individual Scales

As shown in Table 7.13, three studies regressed MMPI scales on some measure of performance. Each of the three studies found that a linear combination of MMPI scales significantly predicted performance. The problem in interpreting these findings is that the same scales did not predict in each study. Such inconsistency is not surprising when entering a large number of scales into a regression using a relatively small sample size.

Regression of Individual Items

Three studies (Corey, 1988; Levine, 1979; Merian, Stefan, Schoenfeld, & Kobos, 1980) attempted to predict performance by entering each of the over 300 individual MMPI items into a regression to predict police performance. Because such attempts are not statistically prudent and lack any basis in theory, they are not recommended as a selection method and will not be discussed further.

Table 7.12: Meta-analysis results for the validity of psychologists' interpretations of personality inventories

Criterion/Inventory	K	N	r	95% Confidence Interval		ρ	90% Credibility Interval		Var	Q_w
				Lower	Upper		Lower	Upper		
Job Performance										
MMPI	11	2,917	.13	.07	.19	.22	.04	.40	54%	20.42*
IPI	7	1,885	.18	.09	.26	.29	.04	.55	42%	16.67*
Inwald studies	4	1,265	.24	.19	.30	.40	.40	.40	100%	1.35
Non-Inwald studies	3	620	.04	- .08	.15	- .13	- .13	.25	51%	5.88
CPI	2	204	.21	.07	.34	.34	.34	.34	100%	0.36
MMPI + IPI	7	2,074	.22	.12	.33	.07	.07	.67	38%	18.48*
Inwald studies	5	1,484	.25	.20	.38	.37	.37	.58	86%	5.80
Non-Inwald studies	2	590	.06	- .02	.14	.10	.10	.10	100%	0.05
MMPI + CPI	3	423	.17	- .02	.36	- .09	- .09	.66	33%	9.08*
Academy Performance										
MMPI	4	768	.21	.15	.28	.31	.31	.31	100%	2.96
IPI	2	496	.22	.14	.31	.32	.32	.32	100%	0.77
MMPI + IPI	2	496	.30	.22	.38	.43	.43	.43	100%	0.86
MMPI + CPI	3	248	.29	.17	.40	.41	.41	.41	100%	0.26

K=number of studies, N=sample size, r = mean correlation, ρ = mean correlation corrected for range restriction, criterion unreliability, and predictor reliability, VAR = percentage of variance explained by sampling error and study artifacts, Q_w = the within group heterogeneity

Table 7.13: Meta-analysis of studies using multiple regression of scale scores

Criterion/Inventory	K	N	r	95% Confidence Interval		ρ	90% Credibility Interval		Var	Q_w
				Lower	Upper		Lower	Upper		
Job Performance										
MMPI	3	436	.28	.15	.41	.46	.36	.55	93%	3.29
IPI	2	301	.37	.09	.65	.60	.19	.99	41%	4.90*
CPI	2	226	.43	.32	.53	.67	.67	.67	100%	0.02
MMPI + IPI	1	219	.19							

K=number of studies, N=sample size, r = mean correlation, ρ = mean correlation corrected for range restriction, criterion unreliability, and predictor reliability, VAR = percentage of variance explained by sampling error and study artifacts, Q_w = the within group heterogeneity

Psychopathology Categories

The previous discussion focused on the validity of specific personality inventories. Another way to look at the validity of personality inventories is to combine results across inventories and focus on broader constructs. To do so, scales from personality inventories focusing on psychopathology were grouped by the specific types of disorders they purport to measure (e.g., depression, schizophrenia). A summary of these groupings is shown in Appendix 7A. As shown in Tables 7.14 and 7.15, grouping the individual scales into categories of psychopathology did little to change the validity of individual scales. The only psychopathology constructs with significant validity (confidence interval did not include zero) that could be generalized across studies (at least 75% of the variance due to sampling and measurement error) were substance abuse as a predictor of disciplinary problems and family problems as a predictor of performance ratings.

Table 7.14: Meta-analysis of psychopathology factors in predicting supervisor ratings of performance

Psychopathology Factor	K	N	r	95% Confidence Interval		ρ	90% Credibility Interval		Var	Q_w
				Lower	Upper		Lower	Upper		
Test Taking Style										
Defensiveness	58	8,292	- .02	- .09	.05	- .04	- .43	.34	40%	146.34*
Malingering	23	3,304	- .09	- .15	- .02	- .15	- .45	.15	49%	46.74*
Psychopathology Construct										
Hypochondriasis	30	4,020	- .05	- .11	.01	- .08	- .36	.20	54%	55.47*
Depression	31	3,795	- .08	- .13	- .03	- .14	- .31	.03	71%	43.41
Hysteria	28	4,456	- .01	- .08	.05	- .02	- .29	.24	43%	64.61*
Antisocial/aggressive	30	4,135	- .10	- .17	- .03	- .19	- .55	.18	46%	65.70*
Paranoia	34	4,772	- .04	- .12	.04	- .08	- .52	.36	31%	100.78*
Psychasthenia	28	3,946	- .08	- .13	- .03	- .15	- .30	.00	74%	36.78
Schizophrenia	29	3,718	- .09	- .18	- .01	- .18	- .63	.28	35%	81.79*
Mania	29	3,969	- .11	- .17	- .04	- .20	- .51	.11	54%	53.75*
Social introversion	30	3,872	- .04	- .11	.02	- 08	- .40	.24	51%	58.41*
Anxiety	5	811	- .09	- .16	- .02	- .15	- .15	- .15	100%	0.61
Substance abuse	8	1,645	- .07	- .15	.01	- .12	- .43	.19	33%	27.22*
Family problems	9	1,433	- .13	- .20	- .07	- .23	- .37	- .08	74%	12.13
Sexual concerns	4	668	- .13	- .21	- .06	- .22	- .22	- .22	100%	3.74

Table 7.15: Meta-analysis of psychopathology factors in predicting disciplinary problems

Psychopathology Factor	K	N	r	95% Confidence Interval		ρ	90% Credibility Interval		Var	Q_w
				Lower	Upper		Lower	Upper		
Test Taking Style										
Defensiveness	24	9,505	- .01	- .05	.02	- .02	- .16	.11	58%	41.34*
Malingering	11	4,366	.02	- .02	.06	.03	- .11	.17	63%	17.34
Psychopathology Construct										
Hypochondriasis	13	3,814	- .02	- .06	.02	- .03	- .16	.10	72%	18.12
Depression	14	4,823	- .01	- .06	.03	- .02	- .22	.17	51%	27.66*
Hysteria	14	5,087	- .01	- .05	.03	- .02	- .22	.18	48%	28.94*
Antisocial/aggressive	19	6,236	.04	- .01	.09	.07	- .15	.30	44%	42.84*
Paranoia	13	4,490	.02	- .02	.07	.05	- .14	.23	54%	24.23*
Psychasthenia	16	4,547	- .04	- .07	.00	- .07	- .21	.07	71%	22.68
Schizophrenia	16	5,005	.01	- .02	.03	.01	- .11	.13	75%	21.35
Mania	15	4,863	.02	- .01	.05	.04	- .04	.12	87%	17.25
Social introversion	17	5,005	.01	- .01	.04	.03	- .08	.14	73%	23.31
Borderline	2	1,111	.00	- .06	.06	.00	- .15	.16	42%	4.79*
Anxiety	3	1,857	- .01	- .06	.04	- .01	- .11	.09	59%	5.09
Substance abuse	8	3,102	.05	.01	.08	.09	.09	.09	100%	4.61

Big-5 Personality Dimensions

A common practice in conducting meta-analyses of personality scores is to place the dimensions from each personality inventory into one of the Big 5 personality factors (openness to experience, conscientiousness, extraversion, agreeableness, emotional stability). Appendix 7B lists the personality scales used in this meta-analysis and their respective Big 5 category. Only personality inventories designed to measure normal personality were included in this analysis. Inventories such as the MMPI, IPI, and MCMI were not included as they are designed to measure psychopathology.

Scales were placed in the Big 5 categories on the basis of their scale descriptions and their correlations with tests tapping Big 5 dimensions. The accuracy of this process was checked by comparing my categories with those used by Ones (1993). There were only a few scales that I had not placed in the same category as Ones (1993). To solve these few discrepancies, I had two psychologists look at the scales and determine which of the categories they thought was most correct.

As shown in Table 7.16, the Big 5 factors were generally able to predict academy and patrol performance. Openness was the best predictor of academy grades, conscientiousness was the best predictor of supervisor ratings of performance, and emotional stability was the best predictor of discipline problems. It should be noted, however, that the individual scales of the CPI were better predictors of the criteria than were the Big 5 factors. This information, combined with the high degree of variability within each factor, suggests that the Big 5 factors may be too broad, and we would be better served sticking with individual scales.

Table 7.16: Meta-analysis of Big 5 personality factors (Inventoris of normal personality)

Criterion/Big 5 Factor	K	N	r	95% Confidence Interval		ρ	90% Credibility Interval		Var	Q_w
				Lower	Upper		Lower	Upper		
Academy Grades										
Openness to Experience	19	2,317	.22	.14	.29	.34	.01	.68	38%	49.80*
Conscientiousness	15	2,652	.13	.06	.21	.21	- .10	.53	32%	47.50*
Extraversion	34	5,515	.12	.08	.16	.19	- .02	.41	50%	67.52*
Agreeableness	14	3,880	.04	.00	.08	.06	- .06	.18	65%	21.56
Emotional stability	13	3,085	.11	.05	.16	.17	.00	.34	54%	24.25*
Performance Ratings										
Openness to Experience	71	5,975	.08	.04	.12	.14	- .22	.50	46%	154.08*
Conscientiousness	68	6,818	.12	.09	.15	.22	- .02	.45	63%	100.55*
Extraversion	84	9,984	.05	.03	.07	.09	- .01	.18	89%	94.23
Agreeableness	69	7,653	.07	.04	.10	.13	- .07	.33	67%	102.24*
Emotional stability	50	5,772	.09	.06	.13	.17	- .03	.37	68%	73.07*
Discipline Problems										
Openness to Experience	17	3,338	- .07	- .12	- .02	- .13	- .34	.08	51%	33.15*
Conscientiousness	22	5,823	- .07	- .11	- .04	- .13	- .30	.04	54%	40.73*
Extraversion	32	10,887	.01	- .02	.03	.01	- .12	.14	60%	53.62*
Agreeableness	25	10,449	- .06	- .08	- .03	- .10	- .22	.01	61%	40.77*
Emotional stability	26	8,805	- .09	- .12	- .05	- .15	- .36	.05	40%	64.56*

Personality Profiles

MMPI

Many of the studies provided mean test scores for the personality test being studied. To compute a "police profile" for these tests, the means from the studies were combined. As can be seen from Table 7.17, there are big differences in the profiles formed from the MMPI and MMPI-2. On the original MMPI, police officers have elevated scores on the K, Pd, and Ma scales, indicating that as a group they took the test in a defensive manner, are rebellious, and energetic. On the MMPI-2, however, the only significant elevations are in the L and K scales. Such a profile would indicate defensiveness when taking the test, but an otherwise normal profile with no other scale elevations.

There are three potential reasons for the differences in profiles. One reason could be that the differences in scores represent actual test and norm differences between the MMPI and MMPI-2. A second reason could be that because the MMPI-2 was given to applicants beginning in the late 1990s, the profile difference could be a function of personality differences between recent applicants and their predecessors from previous decades. A third reason could be testing changes brought about by the Americans with Disabilities Act (ADA). Prior to 1990, the MMPI was routinely administered to all applicants. After the passage of the ADA, however, tests of psychopathology could only be administered after a conditional offer of employment. Thus, studies using the MMPI-2 were based on applicants who had already passed interviews, cognitive ability tests, and a background check.

As shown in Table 7.18, the only noticeable differences between the applicant and incumbent samples were that incumbents showed less defensiveness and scored lower on femininity than the applicants.

Table 7.17 Law Enforcement Norms for the MMPI and MMPI-2

Scale	All Studies			MMPI Version					
				MMPI			MMPI-2		
	K	N	Mean	K	N	Mean	K	N	Mean
L	96	15,501	53.0	78	10,873	52.4	18	4,628	58.9
F	101	16,554	48.2	80	11,657	49.7	21	4,887	44.8
K	96	15,566	59.6	78	10,938	59.8	18	4,628	58.7
Hs	102	15,619	49.6	83	11,206	49.8	19	4,413	49.2
D	102	15,848	50.2	82	11,116	51.2	20	4,732	47.6
Hy	101	15,529	53.2	82	11,116	55.0	19	4,413	48.9
Pd	105	16,464	55.9	84	11,598	57.9	21	4,866	52.1
Mf	97	15,368	51.6	78	11,041	54.7	19	4,327	43.8
Pa	102	15,848	50.0	82	11,116	51.7	20	4,732	46.1
Pt	102	15,619	50.9	83	11,206	52.2	19	4,413	47.5
Sc	103	15,873	51.4	83	11,141	52.8	20	4,732	48.2
Ma	103	15,688	55.0	83	11,141	56.5	20	4,547	50.7
Si	100	15,268	44.9	81	10,855	45.6	19	4,413	43.2

Table 7.18 MMPI Law Enforcement Norms for Applicants and Incumbents

Scale	All Studies			Sample					
				Applicants			Incumbents		
	K	N	Mean	K	N	Mean	K	N	Mean
L	96	15,501	53.0	24	3,376	52.9	72	12,125	54.8
F	101	16,554	48.2	24	3,376	49.7	77	13,178	48.0
K	96	15,566	59.6	24	3,376	61.5	72	12,190	58.9
Hs	102	15,619	49.6	24	3,376	49.8	78	12,243	49.6
D	102	15,848	50.2	24	3,376	50.8	78	12,472	50.0
Hy	101	15,529	53.2	24	3,376	54.7	77	12,153	52.8
Pd	105	16,464	55.9	25	3,510	57.9	80	12,954	55.4
Mf	97	15,368	51.6	24	3,290	55.1	73	12,078	50.7
Pa	102	15,848	50.0	24	3,376	52.2	78	12,472	49.4
Pt	102	15,619	50.9	24	3,376	52.2	78	12,243	50.5
Sc	103	15,873	51.4	24	3,376	52.9	79	12,497	51.1
Ma	103	15,688	55.0	25	3,510	55.8	78	12,178	54.6
Si	100	15,268	44.9	24	3,376	44.3	76	11,892	45.0

Table 7.19 CPI Law Enforcement Norms

Scale	K	N	Mean
Dominance	31	3,373	57.4
Capacity for status	33	3,475	52.5
Sociability	28	3,101	54.4
Social presence	30	3,285	55.9
Self-acceptance	30	3,285	56.2
Well-being	33	3,477	54.4
Responsibility	33	3,477	50.6
Socialization	30	3,285	53.3
Self-control	33	3,397	54.8
Tolerance	33	3,397	52.9
Good impression	33	3,397	55.4
Communality	33	3,397	55.2
Achievement via conformance	33	3,397	57.3
Achievement via independence	31	3,295	55.9
Intellectual efficiency	31	3,295	53.4
Psychological mindedness	31	3,295	56.8
Flexibility	30	3,205	50.6
Femininity	30	3,205	46.7

CPI

As shown in Table 7.19, the CPI profile for law enforcement personnel represents a person who is functioning effectively both socially and intellectually. The highest scores were on the dominance and achievement via conformance scales and the lowest were on the femininity, flexibility, and responsibility scales.

Appendix 7A
Reliabilities and Descriptions of Personality Inventories Measuring Psychopathology

Test/Scale	Reliability		Psychopathology Category	Description of high scorers
	T-R	Internal		
MMPI-2 (Graham, 1993)				
L	.77	.62	Defensiveness	Defensive, conventional
F	.78	.64	Malingering	Moody, restless, troubled
K	.84	.74	Defensiveness	Defensive, inhibited
Hs	.85	.77	Hypochondriasis	Complains, pessimistic
D	.75	.59	Depression	Depressed, sad, helpless
Hy	.72	.58	Hysteria	Immature, self-centered
Pd	.81	.60	Antisocial	Rebellious, shallow
Mf	.82	.58		Artistic, sensitive, feminine
Pa	.67	.34	Paranoia	Suspicious, hostile, paranoid
Pt	.89	.85	Psychasthenia	Anxious, neat, moral, rigid
Sc	.87	.85	Schizophrenia	Stubborn, unusual
Ma	.83	.58	Mania	Energetic, talkative, confident
Si	.92	.82	Social introversion	Introverted, sensitive, moody
Personality Assessment Inventory (Morey, 1991)				
Inconsistency	.31	.45		Inconsistent responding
Infrequency	.48	.52		Careless or random responding
Negative impression	.75	.72	Malingering	Exaggerating symptoms
Positive impression	.78	.71	Defensiveness	Faking good or reluctant to admit flaws
Somatic complaints	.83	.89	Hypochondriasis	Preoccupation with health matters
Anxiety	.88	.90	Anxiety	Tension and negative affect
Anxiety-related disorders	.83	.76	Anxiety-related disorders	Phobias, traumatic stress, obsessive-compulsive
Depression	.87	.87	Depression	Cognitive and affective depression

Test/Scale	Reliability		Psychopathology Category	Description of high scorers
	T-R	Internal		
Mania	.83	.82	Mania	High activity levels, feelings of grandiosity
Paranoia	.84	.85	Paranoia	Suspiciousness, feelings of persecution
Schizophrenia	.82	.81	Schizophrenia	Thought disorders and psychotic experiences
Borderline features	.86	.87	Borderline	Impulsive, unstable interpersonal relations
Antisocial features	.89	.84	Antisocial	Problems with authority, lacks empathy
Alcohol problems	.92	.84	Substance abuse	Problems with alcohol
Drug problems	.79	.74	Substance abuse	Problems with drugs
Aggression	.85	.85	Antisocial	Aggressive, angry, hostile
Suicidal ideations	.71	.85	Depression	Thinks about committing suicide
Stress	.88	.76	Anxiety	Feels stressed and that life is in turmoil
Non-support	.81	.72		Perceives a lack of social support
Treatment rejection	.83	.76		Low motivation for treatment
Dominance	.77	.78	Big 5: Extraversion	Self-assured, confident, dominant
Warmth	.74	.79	Big 5: Agreeableness	Warm, friendly, sympathetic
Millon Clinical Multiaxial Inventory				
Disclosure			Defensiveness	
Desirability			Defensiveness	
Debasement	.82		Malingering	Poor self-image, resents others
Schizoid	.71		Social introversion	Socially detached, prefers being alone
Avoidant	.76		Social introversion	Anxious in social settings
Depressive			Depression	Pessimistic outlook in life, downcast
Dependent	.67		Dependent	Submissive, agreeable, passive
Histrionic	.80		Histrionic	Needs to be the center of attention

Test/Scale	Reliability		Psychopathology Category	Description of High Scorers
	T-R	Internal		
Narcissistic	.87		Narcissistic	Self-centered, arrogant
Antisocial	.85		Antisocial	Irresponsible, independent
Sadistic			Antisocial	Aggressive and hostile toward others
Compulsive	.44	.66	Psychasthenia	Perfectionistic, organized, orderly
Passive-aggressive			Passive-aggressive	Argumentative, has a negative attitude
Self-defeating			Negativistic	Is taken advantage of, self-sacrificing
Schizotypal	.75		Social introversion	Fears human contact, distrusts others
Borderline	.64		Borderline	Instability in moods and self-image
Paranoid			Paranoia	Defensive, suspicious, distrusting
Anxiety disorder			Anxiety	Tense, on edge, anxious
Somatoform disorder	.90		Hysteria	Complains about vague physical problems
Bipolar			Mania	Energetic, impulsive, restless
Dysthymia			Depression	Apathetic, tired, discouraged
Alcohol dependence			Substance abuse	Abuses alcohol
Drug dependence			Substance abuse	Abuses drugs
Post-traumatic stress			Anxiety-related disorders	Anxious, suffers from intrusive memories
Thought disorders			Schizophrenia	Suffers from hallucinations and odd thinking
Major depression			Depression	Severely depressed, can't function properly
Delusional			Paranoia	Feels persecuted, has delusional thinking

Appendix 7B
Personality Scale Reliabilities and Descriptions

Test/Scale	Reliability		Big 5 Category	Description of High Scorers
	T-R	Internal		
California Psychological Inventory (Gough & Bradley, 1996)				
Dominance	.72	.83	Extraversion	Aggressive, persuasive, confident
Capacity for status	.68	.72	Extraversion*	Ambitious, active, forceful
Sociability	.71	.77	Extraversion	Outgoing, sociable, participative
Social presence	.63	.71	Extraversion	Poised, spontaneous, confident
Self-acceptance	.71	.67	Openness	Outspoken, intelligent, confident
Well being	.72	.84	Stability	Energetic, ambitious, versatile
Responsibility	.73	.77	Conscientiousness	Conscientious, responsible, dependable
Socialization	.69	.78	Agreeableness*	Serious, honest, modest, sincere
Self-control	.75	.83	Stability	Calm, patient, practical, deliberate
Tolerance	.71	.79	Openness	Tolerant, non-judgmental, resourceful
Good impression	.69	.81	Agreeableness	Cooperative, warm, helpful
Communality	.44	.71	Agreeableness	Sincere, patient, tactful, reliable
Ach via conf	.73	.78	Conscientiousness	Organized, responsible, cooperative
Ach via indep	.63	.80	Conscientiousness	Mature, self-reliant, demanding
Intellectual eff	.77	.79	Openness	Intelligent, capable, clear thinking
Psych mindedness	.49	.62	Stability	Perceptive, spontaneous, observant
Flexibility	.60	.64	Openness	Flexible, adaptable, informal
Femininity	.65	.73	Agreeableness	Appreciative, sincere, helpful
V1	.72	.82	Extraversion (-)	Introverted,
V2	.70	.77	Conscientiousness	Orderly, conforming
V3	.74	.88	Agreeableness	Trusting, stable, agreeable

Mgt potential	.65	.81	Stability	
Work orientation	.67	.78	Conscientiousness	Dependable, conscientious, organized
Creative temp	.65	.73	Openness	Adventurous, clever, unconventional
Leadership	.73	.88	Extraversion	Alert, ambitious, energetic, poised
Amicability	.70	.79	Agreeableness	Modest, reasonable, tactful
Law enforcement	.56	.45	Conscientiousness	Confident, organized, practical
Tough-minded	.65	.79	Stability	Practical, logical, unemotional
Sixteen Personality Factor Questionnaire (16PF)				
A: Outgoing	.82	.90	Extraversion	Warm, sociable, trustful, caring
B: Bright	.65	.86	Openness	Intellectual, cultured, quick
C: Calm	.79	.93	Stability	Mature, calm, stable, even-tempered
E: Dominant	.80	.91	Extraversion	Assertive, unconventional, competitive
F: Happiness	.81	.84	Extraversion	Talkative, enthusiastic, cheerful
G: Conscientious	.81	.85	Conscientiousness	Conscientious, persistent, responsible
H: Venturesome	.87	.83	Extraversion	Adventurous, friendly, carefree, talkative
I: Tender-minded	.83	.76	Agreeableness	Sensitive, kind, gentle, helpful
L: Suspicious	.75	.77	Agreeableness (-)	Suspicious, hard, irritable, jealous
M: Imaginative	.70	.88	Openness	Imaginative, creative, artistic
N: Shrewd	.67	.79	Extraversion	Insightful, ambitious, polished
O: Apprehensive	.79	.85	Stability (-)	Depressed, anxious, fussy
Q1: Experiments	.78	.71	Openness	Open to change, experimenting
Q2: Self-directed	.76	.79	Openness	Self-reliant, self-sufficient
Q3: Disciplined	.78	.76	Conscientiousness	Organized, reliable, conscientious
Q4: Tense	.82	.88	Stability (-)	Tense, impatient, fast-paced

Employee Personality Inventory (EPI)				
Thinking	.79		Openness	Smart, adaptable, original
Directing	.85		Extraversion	Assertive, persuasive, confident
Communicating	.85		Extraversion	Outgoing, friendly, talkative
Soothing	.90		Agreeableness	Calm, loyal, sincere
Organizing	.84		Conscientiousness	Organized, detailed, responsible
Guilford-Zimmerman Temperament Survey (Guilford et al., 1976)				
General activity	.67	.79	Extraversion	Energetic, enthusiastic
Restraint	.74	.80	Conscientiousness	Deliberate, restrained, self-controlled
Ascendance	.53	.82	Extraversion	Enjoys leading, speaking with others, persuading
Sociability	.71	.87	Extraversion	Has many friends, likes social activities
Stability	.71	.84	Stability	Optimistic, even-tempered
Objectivity	.64	.75	Stability	Thick-skinned, stays out of trouble
Friendliness	.65	.75	Agreeableness	Tolerates others, respects others
Thoughtfulness	.58	.80	Openness	Mentally poised, interested in thinking
Personal relations	.64	.80	Agreeableness	Tolerant, has faith in others
Masculinity	.80	.85		Fearless, interested in masculine activities
Edwards Personal Preference System (EPPS; Edwards, 1959)				
Achievement	.74	.74	Conscientiousness	Achievement oriented, motivated
Deference	.78	.60	Conscientiousness	Follows, rule oriented, praises others
Order	.87	.74	Conscientiousness	Neat, organized, makes plans
Exhibition	.74	.61	Extraversion	Witty, tells jokes and stories
Autonomy	.83	.76	Openness	Independent, unconventional
Affiliation	.77	.70	Agreeableness	Loyal, sharing, group oriented
Intraception	.86	.79	Agreeableness	Empathic, introspective
Succorance	.78	.76	Stability (-)	High need for the support of others
Dominance	.87	.81	Extraversion	Leader, assertive, persuasive

Abasement	.88	.84	Stability	Feel guilty, accept blame
Nurturance	.79	.78	Agreeableness	Helpful, kind, generous
Change	.83	.79	Openness	Experimenting, tries new things
Endurance	.86	.81	Conscientiousness	Persistent, hard working
Heterosexuality	.85	.87		Heterosexual
Aggression	.78	.84	Stability (-)	Aggressive, angry, critical
Personal Perspectives Inventory (PPI)				
Openness	.87	.84	Openness	Bright, inquisitive, open-minded
Conscientiousness	.92	.88	Conscientiousness	Dependable, reliable, motivated
Extraversion	.89	.79	Extraversion	Outgoing, sociable, talkative
Agreeableness	.90	.78	Agreeableness	Kind, helpful, loyal
Stability	.86	.85	Stability	Calm, stable
NEO-PI (Caruso, 2000)				
Openness	.78	.79	Openness	Bright, inquisitive, open-minded
Conscientiousness	.76	.84	Conscientiousness	Dependable, reliable, motivated
Extraversion	.81	.83	Extraversion	Outgoing, sociable, talkative, energetic
Agreeableness	.58	.76	Agreeableness	Kind, helpful, loyal
Neuroticism	.82	.88	Stability	Calm, stable

Chapter References

*Aadland, R. L. (1981). *The prediction of use of deadly force by police officers in simulated field situations.* Unpublished doctoral dissertation, California School of Professional Psychology, Los Angeles.

*Aamodt, M. G., & Kimbrough, W. W. (1985). Personality differences between police and fire applicants. *Journal of Police and Criminal Psychology, 1*(1), 10-13.

*Abraham, J. D., & Morrison, J. D. (2003). Relationship between the Performance Perspectives Inventory's Conscientiousness scale and job performance of corporate security guards. *Applied H.R.M. Research*, 8(1), 45-48.

*Allison, D. A. (1991). *The relationship between gender, ethnicity, age, and personality traits among police officers*. Unpublished doctoral dissertation, California School of Professional Psychology.

*Azen, S. P., Snibbe, H. M., & Montgomery, H. R. (1973). A longitudinal predictive study of success and performance of law enforcement officers. *Journal of Applied Psychology, 57*(2), 190-192.

*Azen, S. P., Snibbe, H. M., Montgomery, H. R., Fabricatore, J., & Earle, H. (1974). A longitudinal predictive study of success and performance of law enforcement officers. *American Journal of Community Psychology, 1*(2), 79-86.

*Balch, D. E. (1977). *Personality trait differences between successful and non-successful police recruits at a typical police academy and veteran police officers.* Unpublished doctoral dissertation, United States International University.

*Band, S. R., & Manuele, C. A. (1987). Stress and police officer performance: An examination of effective coping behavior. *Journal of Police and Criminal Psychology, 3*(3), 30-42.

*Banks, D. E. (1988). *The relationship of personality styles to police job performance*. Unpublished doctoral dissertation, California School of Professional Psychology.

*Bartol, C. R. (1991). Predictive validation of the MMPI for small-town police officers who fail. *Professional Psychology: Research and Practice, 22*(2), 127-132.

*Bartol, C. R. (1982). Psychological characteristics of small-town police officers. *Journal of Police Science and Administration, 10(1)*, 58-63.

*Bartol, C. R., Bergen, G. T., Volckens, J. S., & Knoras, K. M. (1992). Women in small-town policing. *Criminal Justice and Behavior, 19(3)*, 240-259.

*Benner, A. W. (1991). *The changing cop: A longitudinal study of psychological testing within law enforcement.* Unpublished doctoral dissertation, Saybrook Institute (now called the Saybrook Graduate School in San Francisco).

*Bernstein, I. H., Schoenfeld, L. S., & Costello, R. M. (1982). Truncated component regression, multicollinearity and the MMPI's use in a police officer selection setting. *Multivariate Behavioral Research, 17*, 99-116.

*Beutler, L., Storm, A., Kirkish, P., Scogin, F., & Gaines, J. A. (1985). Parameters in the prediction of police officer performance. *Professional Psychology: Research and Practice, 16*(2), 324-335.

*Beutler, L.E., Nussbaum, P. D., & Meredith, K. E. (1988). Changing personality patterns of police officers. *Professional Psychology: Research and Practice, 19*(5), 503-507.

*Black, J. (2000). Personality testing and police selection: Utility of the "Big Five." *New Zealand Journal of Psychology, 29*(1), 2-9.

*Blau, T. H., Super, J. T., & Brady, L. (1993). The MMPI good cop/bad cop profile in identifying dysfunctional law enforcement personnel. *Journal of Police and Criminal Psychology, 9*(1), 2-4.

*Blunt, J. H. (1982). *The prediction of police officer performance utilizing the MMPI.* Unpublished master's thesis, University of Central Florida

*Boes, J. O., Chandler, C. J., & Timm, H. W. (1997). *Police integrity: Use of personality measures to identify corruption-prone officers.* Monterey, CA: Defense Personnel Security Research Center.

*Borum, R. (1991) *The MMPI and IPI as predictors of suitability of law enforcement applicants.* Unpublished doctoral dissertation, Florida Institute of Technology.

*Bowen, D. N. (1984). *Personality and demographic characteristics of road deputies and correctional officers.* Unpublished doctoral dissertation, Florida Institute of Technology.

*Boyce, T. N. (1988). *Psychological screening for high-risk police specialization.* Unpublished doctoral dissertation, Georgia State University.

*Bozza, C. M. (1990). *Improving the prediction of police officer performance from screening information.* Unpublished doctoral dissertation, United States International University.

*Bradford, A. C. (1991). *Psychological screening for narcotics officers and detectives.* Unpublished doctoral dissertation, Miami University of Ohio.

*Brewster, J. (1996). Hypervigilance and cynicism in police officers. *Journal of Police and Criminal Psychology, 10*(4), 7-9.

*Brewster, J., & Stoloff, M. L. (1999). Using the good cop/bad cop profile with the MMPI-2. *Journal of Police and Criminal Psychology, 14*(2), 29-34.

*Brown, G. V. (1996). *Acceptable vs. marginal police officers' psychological ratings: A longitudinal comparison of job performance.* Unpublished doctoral dissertation, Florida International University.

*Caldwell-Andrews, A. A. (2000). *Relationships between MMPI-2 validity scales and NEO PI-R experimental validity scales in police candidates.* Unpublished doctoral dissertation, University of Kentucky.

Caruso, J. C. (2000). Reliability generalization of the NEO personality scales. *Educational and Psychological Measurement, 60*, 236-254.

*Cauthen, D. (1987). *Urban police applicant MMPI score differences due to employment classification and gender.* Unpublished doctoral dissertation, Oklahoma State University.

*Clopton, W. (1971). *Comparison of ratings and field performance data in validating predictions of patrolman performance: A five-year follow-up study.* Unpublished master's thesis, University of Cincinnati.

*Cope, J. R. (1981). *Personality characteristics of successful versus unsuccessful police officers.* Unpublished doctoral dissertation, Florida Institute of Technology.

*Corey D.M. (1988). *The psychological suitability of police officer candidates.* Unpublished doctoral dissertation, The Fielding Institute.

*Cortina, J. M., Doherty, M. L, Schmitt, N., Kaufman, G., & Smith, R. G. (1992). The "Big Five" personality factors in the IPI and MMPI: Predictors of police performance. *Personnel Psychology, 45,* 119-140.

*Cortina, J. M., Doherty, M. L, Schmitt, N., Kaufman, G., & Smith, R. G. (1991). *Validation of the IPI and MMPI as predictors of*

police performance. Paper presented at the annual meeting of the Society for Industrial-Organizational Psychology.
*Costello, R. M., & Schneider, S. L. (1996). Validation of a preemployment MMPI index correlated with disciplinary suspension days of police officers. *Psychology, Crime & Law, 2*, 299-306.
*Costello, R. M., Schneider, S. L. & Schoenfeld, L. S. (1993). Applicants' fraud in law enforcement. *Psychological Reports, 73*(1), 179-183.
*Costello, R. M., & Schoenfeld, L. S. (1981). Time-related effects on MMPI profiles of police academy recruits. *Journal of Clinical Psychology, 37*(3), 518-522.
*Costello, R. M., Schneider, Schoenfeld, L. S., & Kobos, J. (1982). Police applicant screening: An analogue study. *Journal of Clinical Psychology, 38*(1), 216-221.
*Cottle, H. D., & Ford, G. G. (2000). The effects of tenure on police officer personality. *Journal of Police and Criminal Psychology, 15*(1), 1-9
*Cowan, L. B. (1991). *MMPI performance related to length of service for public safety employees.* Unpublished doctoral dissertation, Purdue University.
*Daley, R. E. (1978). *The relationship of personality variables to suitability for police work* Unpublished doctoral dissertation, Florida Institute of Technology.
*Dantzker, M. L., & Freeberg, D. (2003). An exploratory examination of pre-employment psychological testing of police officer candidates with a Hispanic surname. *Journal of Police and Criminal Psychology, 18*(1), 38-44.
*Davidson, N. B. (1975). *The predictive validity of a police officer selection program.* Unpublished master's thesis, Portland State University.
*Davis, R. D., Rostow, C. D., Pinkston, J. B., & Cowick, L. M. (1999). An investigation into the usefulness of the MMPI and MMPI-2 in municipal and state police candidate selection. *Journal of Police and Criminal Psychology, 14*(1), 100-106.
*Dean, D. (1974). *The relationship between Eysenckian personality variables and ratings of job performance and promotion potential of a group of police officers.* Unpublished doctoral dissertation, Ball State University.
*Detrick, P., Chibnall, J. T., & Rosso, M. (2001). Minnesota Multiphasic Personality Inventory-2 in police officer selection: Normative

data and relation to the Inwald Personality Inventory, *Professional Psychology: Research and Practice, 32*(5), 484-490.

*Dorner, K. R. (1991). *Personality characteristics and demographic variables as predictors of job performance in female traffic officers.* Unpublished doctoral dissertation, United States International University.

Edwards, A. L. (1959). *Edwards Personal Preference Schedule.* New York: Psychological Corporation.

*Elam, J. D. (1983). *Minnesota Multiphasic Personality Inventory and California Psychological Inventory as predictors of performance for a municipal and a state police agency.* Unpublished doctoral dissertation, University of Oklahoma.

*Fabricatore, J., Azen, S., Schoentgen, S., & Snibbe, H. (1978). Predicting performance of police officers using the Sixteen Personality Factor Questionnaire. *American Journal of Community Psychology, 6*(1), 63-69.

*FitzGerald, P. R. (1986). *The prediction of police performance using the MMPI and CPI.* Unpublished doctoral dissertation, Saint Louis University.

*Forschner, B. E. (1981). *An analysis of the personality characteristics of undergraduate criminal justice majors and their field counterparts.* Unpublished doctoral dissertation, Ohio State University.

*Funk, A. P. (1997). *Psychological assessment of military federal agents using the MMPI-2: A closer look at employment selection and performance prediction.* Unpublished master's thesis, Florida State University.

*Fuqua, D. R. (1975). *A comparison of police and criminal personality characteristics as measured by the MMPI.* Unpublished master's thesis, Eastern Illinois University.

*Gardner, J. F. (1994). *The predictive validity of psychological testing in law enforcement.* Unpublished master's thesis, University of Alabama.

*Gelbart, M. (1978). *Psychological, personality, and biographical variables related to success as a hostage negotiator.* Unpublished doctoral dissertation, University of Southern California.

*Geraghty, M. F. (1986). *The California Personality Inventory test as a predictor of law enforcement officer job performance.*

Unpublished doctoral dissertation, Florida Institute of Technology.

*Gonder, M. L. (1998). *Personality profiles of police officers: Differences in those that complete and fail to complete a police training academy.* Unpublished master's thesis, University of North Carolina-Charlotte.

*Gottesman, J. I. (1974). *The utility of the MMPI in assessing the personality patterns of urban police applicants.* Unpublished doctoral dissertation, Stevens Institute of Technology.

*Gottlieb, M. C., & Baker, C. F. (1974). Predicting police officer effectiveness. *The Journal of Forensic Psychology, 6*, 35-46.

*Gough, H. G. (1975). *California Psychological Inventory manual.* Palo Alto, CA: Consulting Psychologists Press.

Gough, H. G., & Bradley, P. (1996). *CPI manual* (3rd Ed.). Palto Alto, CA: Consulting Psychologists Press.

*Gould, L. A. (2000). A longitudinal approach to the study of the police personality: Race/Gender differences. *Journal of Police and Criminal Psychology, 15*(2), 41-51.

*Gould, L. A., & Volbrecht, M. (1999). Personality differences between women police recruits, their male counterparts, and the general female population. *Journal of Police and Criminal Psychology, 14*(1), 1-8.

*Grafton, W. L. (1997). *A descriptive investigation of demographic variables among state troopers, and the relationship between personality profiles and class rank in the Louisiana State Police Academy.* Unpublished doctoral dissertation, University of Southern Mississippi.

Graham, J. R. (1993). *MMPI-2: Assessing personality and psychopathology.* New York: Oxford University Press.

*Grayson, L. J. (1986). *Narcissistic personality stules and their effects on job functioning in police officers.* Unpublished doctoral dissertation, California School of Professional Psychology, Los Angeles.

*Greenberg, B. E., Riggs, M., Bryant, F. B., & Smith, B. D. (2003). Validation of a short aggression inventory for law enforcement. *Journal of Police and Criminal Psychology, 18*(2), 12-19.

*Griffith, T. L. (1991). *Correlates of police and correctional officer performance.* Unpublished doctoral dissertation, Florida State University.

Guilford, J. S., Zimmerman, W. S., & Guilford, J. P. (1976). *The Guilford-Zimmerman Temperament Survey handbook.* San Diego, CA: EdITS.

*Hankey, R. O. (1968). *Personality correlates in a role of authority: The Police.* Unpublished doctoral dissertation, University of Southern California.

*Hankey, R. O., Morman, R. R., Kennedy, P. K., & Heywood, H. L. (1965). TAV selection system and state traffic officer job performance, *Police*, March-April, 10-13.

*Hargrave, G. E. (1985). Using the MMPI and CPI to screen law enforcement applicants: A study of reliability and validity of clinician's decisions. *Journal of Police Science and Administration, 13*(3), 221-224.

*Hargrave, G. E. (1987). Screening law enforcement cadets with the MMPI: An analysis of adverse impact. *Journal of Police and Criminal Psychology, 3*(1), 14-19.

*Hargrave, G. E., & Hiatt, D. (1989). Use of the California Psychological Inventory in law enforcement officer selection. *Journal of Personality Assessment* (2), 267-277.

*Hargrave, G. E., & Hiatt, D. (1987). Law enforcement selection with the interview, MMPI, and CPI: A study of reliability and validity. *Journal of Police Science and Administration, 15*(2), 110-117.

*Hargrave, G. E., Hiatt, D., & Gaffney, T. W. (1986). A comparison of MMPI and CPI test profiles for traffic officers and deputy sheriffs. *Journal of Police Science and Administration, 14*(3), 250-258.

*Hargrave, G. E., Norborg. J. M., & Oldenburg, L. (1986). Differences in entry level test and criterion data for male and female police officers. In Reese, J. T. & Goldstein, H. A. (Eds). *Psychological services for law enforcement*, pp 35-42. Washington, D.C.: U.S. Government Printing Office.

*Hart, R. (1981). *The use of the Clinical Analysis Questionnaire in the selection of police officers: A validation study.* Unpublished doctoral dissertation, Florida State University.

*Henderson, N. D. (1979). Criterion-related validity of personality and aptitude scales: A comparison of validation results under voluntary and actual test conditions. In Charles D. Spielberger (Ed.). *Police Selection and Evaluation: Issues and Techniques.* New York: Praeger Publishers.

*Hess, L. R. (1972). *Police entry tests and their predictability of score in police academy and subsequent job performance.* Unpublished doctoral dissertation, Marquette University.

*Heyer, T. (1998). *A follow-up study of the prediction of police officer performance on psychological evaluation variables.* Unpublished doctoral dissertation, Minnesota School of Professional Psychology.

*Hiatt, D., & Hargrave, G. E. (1988). MMPI profiles of problem police officers. *Journal of Personality Assessment, 52*(4), 722-731.

*Hiatt, D., & Hargrave, G. E. (1988). Predicting job performance with psychological screening. *Journal of Police Science and Administration, 16*, 122-125.

*Hofer, S. M. (1994). *On the structure of personality and the relationship of personality to fluid and crystalized intelligence in adulthood.* Unpublished doctoral dissertation, University of Southern California.

*Hogan, R. (1971). Personality characteristics of highly rated policemen. *Personnel Psychology, 24*, 679-686.

Hogan, R., & Kurtines, W. (1975). Personological correlates of police effectiveness. *The Journal of Psychology, 91*, 289-295.

*Hooke, J. F., & Krauss, H. H. (1971). Personality characteristics of successful police sergeant candidates. *Journal of Criminal Law, Criminology, and Police Science, 62*(1), 104-106.

*Horstman, P. L. (1976). *Assessing the California Psychological Inventory for predicting police performance.* Unpublished doctoral dissertation, University of Oklahoma.

*Hwang, G.S. (1988). *Validity of the California Psychological Inventory for Police Selection.* Unpublished master's thesis, North Texas State University.

*Inwald, R. E. (1988). Five-year follow-up of department terminations as predicted by 16 preemployment psychological indicators. *Journal of Applied Psychology, 73(4)*, 703-710.

*Inwald, R. E., & Brockwell, A. L. (1991). Predicting the performance of government security personnel with the IPI and MMPI. *Journal of Personality Assessment, 56(*3), 522-535.

*Inwald, R. E., & Shusman, E. J. (1984). The IPI and MMPI as predictors of academy performance for police recruits. *Journal of Police Science and Administration, 12*(1), 1-11.

*Inwald, R. E., & Shusman, E. J. (1984). Personality and performance sex differences of law enforcement officer recruits. *Journal of Police Science and Administration, 12*(3), 339-347.

*Kauder, B. S. (1999). *Construct-related evidence of validity for the Inwald Personality Inventory and its usefulness for predicting police officer performance.* Unpublished doctoral dissertation, Pacific University (Forest Grove, OR).

*Keith, J. B. (1993). *Personality characteristics of suburban police recruits and their compatibility with traditional police management theory.* Unpublished master's thesis, Governors State University (University Park, IL).

*Kleiman, L. S. (1978). *Ability and personality factors moderating the relationships of police academy training performance with measures of selection and job performance.* Unpublished doctoral dissertation, University of Tennessee, Knoxville.

*Kleiman, L. S., & Gordon, M. E. (1986). An examination of the relationship between police training academy performance and job performance. *Journal of Police Science and Administration, 14*(4), 293-299.

*Klopsch, J. W. (1983). *Police personality change as measured by the MMPI: A five-year longitudinal study.* Unpublished doctoral dissertation, Fuller Theological Seminary.

*Knights, R. M. (1976). *The relationship between the selection process and on-the-job performance of Albuquerque police officers.* Unpublished doctoral dissertation, University of New Mexico.

*Kornfeld, A. D. (1995). Police officer candidate MMPI-2 performance: Gender, ethnic, and normative factors. *Journal of Clinical Psychology, 51*(4), 536-540.

*Leake, S. A. (1988). *Personal characteristics of peace officers: Model traits, self-selection, and organizational-selection factors in a variety of law enforcement categories*, unpublished doctoral dissertation, University of California at Davis.

*Levine, M. (1979). *Development of an MMPI subscale as an aid in police officer selection.* Unpublished doctoral dissertation, California School of Professional Psychology, Berkeley.

*Lorr, M., & Strack, S. (1994). Personality profiles of police candidates. *Journal of Clinical Psychology, 50*(2), 200-207.

*Mandel, K. (1970). *The predictive validity of on-the-job performance of policemen from recruitment selection information.* Unpublished doctoral dissertation, University of Utah.

*Mass, G. (1980). *Using judgment and personality measures to predict effectiveness in policework: An exploratory validation study.* Unpublished doctoral dissertation, Ohio State University.

*Matarazzo, J. D., Allen, B. V., Saslow, G., & Wiens, A. N. (1964). Characteristics of successful policemen and firemen applicants. *Journal of Applied Psychology, 48*(2), 123-133.

*Matthews, B. L. (1993). *Effects of differential work experience on personality characteristics in police officers and deputies.* Unpublished doctoral dissertation, California Graduate Institute.

*Matyas, G. S. (1980). *The relationship of MMPI and biographical data to police performance.* Unpublished doctoral dissertation, University of Missouri – Columbia.

*McDonough, L. B., & Monahan, J. (1975). The quality control of community caretakers: A study of mental health screening in a sheriff's department. *Community Mental Health Journal, 11*(1), 33-43.

*McEuen, O. L. (1981). *Assessment of some personality traits that show a relationship to academy grades, being dismissed from the department, and work evaluation ratings for police officers in Atlanta, Georgia.* Unpublished doctoral dissertation, The Fielding Institute.

*McQuilkin, J. I., Russell, V. L., Frost, A. G., & Faust, W. R. (1990). Psychological test validity for selecting law enforcement officers. *Journal of Police Science and Administration, 17*(4), 289-294.

*Merian, E. M., Stefan, D., Schoenfeld, L. S., & Kobos, J. C. (1980). Screening of police applicants: A 5-item MMPI research index. *Psychological Reports, 47,*155-158.

*Meunier, G. F., Koontz, T., & Weller, R. (1995). Psychological characteristics of reserve police officers. *Journal of Police and Criminal Psychology, 11*(1), 57-59.

*Mills, A. (1990). *Predicting police performance for differing gender and ethnic groups: A longitudinal study.* Unpublished doctoral dissertation, California School of Professional Psychology.

*Mills, C. J., & Bohannon, W. E. (1980). Personality characteristics of effective state police officers. *Journal of Applied Psychology, 65*(6), 680-684.

Morey, L. C. (1991). *Personality assessment inventory: Professional manual.* Odessa, FL: Psychological Assessment Resources.

*Morman, R. R., Hankey, R. O., Heywood, H. L., & Kennedy, P. K. (1965). Multiple relationships of TAV selection system predictors to state traffic officer performance, *Police*, July-August, 41-44.

*Morman, R. R., Hankey, R. O., Kennedy, P. K., & Heywood, H. L. (1965). Predicting state traffic officer performance with TAV selection system theoretical scoring keys, *Police*, May-June, 70-73.

*Morman, R. R., Hankey, R. O., Heywood, H. L., & Liddle, R. (1966). Predicting state traffic cadet academic performance from theoretical TAV selection system scores, *Police*, July-August, 54-58.

*Morman, R. R., Hankey, R. O., Kennedy, P. K., & Jones, E. M. (1966). Academy achievement of state traffic officer cadets related to TAV selection system plus other variables, *Police*, July-August, 30-34.

*Neal, B. (1986). The K scale (MMPI) and job performance. In Reese, J. T. & Goldstein, H. A. (Eds). *Psychological services for law enforcement*, pp 83-90. Washington, D.C.: U.S. Government Printing Office.

*Nemeth, Y. M. (2001). *Predictive validity of the Clinical Analysis Questionnaire in the pre-employment selection of police officers*. Unpublished master's thesis, University of South Alabama.

*Nowicki, S. (1966). A study of the personality characteristics of successful policemen. *Police, 10*, 39-41.

*Ofton, M. A. (1979). *The relationship between Minnesota Multiphasic Personality Inventory (MMPI) profiles of police recruits and performance ratings in their rookie year.* Unpublished master's thesis, Abilene Christian University.

Ones, D. S. (1993). *The construct validity of integrity tests.* Unpublished doctoral dissertation, University of Iowa.

*Palmatier, J. J. (1996). *The big-five factors and hostility in the MMPI and IPI: Predictors of Michigan State Trooper job performance.* Unpublished doctoral dissertation, Michigan State University.

*Powers, W. P. (1996). *An evaluation of the predictive validity of the MMPI s it relates to identifying police officers prone to engage in the use of excessive force.* Unpublished doctoral dissertation, Adler School of Professional Psychology.

*Pugh, G. (1985). The California Psychological Inventory and police selection. *Journal of Police Science and Administration, 13*(2), 172-177.

*Rand, T. M., & Wagner, E. E. (1973). Correlations between Hand Test variables and patrolman performance. *Psychological Reports, 37,* 477-478.

*Raynes, B. L. (1997). *Predicting difficult employees: the relationship between vocational interest, self-esteem, and problem communication styles*. Unpublished master's thesis, Radford University.

*Reischl, S. R. (1977). *Personality profiles of successful and nonsuccessful police promotional candidates administered the California Psychological Inventory.* Unpublished master's thesis, California State University, Long Beach.

*Rostow, C. D., Davis, R. D., Pinkston, J. B., & Corwick, L. M. (1999). The MMPI-e and satisfactory academy performance: Differences and correlations. *Journal of Police and Criminal Psychology, 14*(2), 35-39.

*Rounds, F. E. (1989). *The relationship of the MMPI and the Wollack Alert/PAF for police applicant selection.* Unpublished master's thesis, East Carolina University.

*Saccuzzo, D. P., Higgins, G., & Lewandowski, D. (1974). Program for psychological assessment of law enforcement officers: Initial evaluation. *Psychological Reports, 35,* 651-654.

*Sarchione, C. D. (1995). *Personality constructs and California Psychological Inventory Subscales as a predictor of job difficulty n police officers.* Unpublished master's thesis, University of North Carolina, Greensboro.

*Sarchione, C. D., Cuttler, M. J., Muchinsky, P. M., & Nelson-Gray, R. O. (1998). Prediction of dysfunctional job behaviors among law enforcement officers. *Journal of Applied Psychology, 83*(6), 904-912

*Saxe, S. J., & Reiser, M. (1976). A comparison of three police applicant groups using the MMPI. *Journal of Police Science and Administration, 4*(4), 419-425.

*Schelling, M. J. (1993). *The use of the Law Enforcement Selection Inventory in the selection of communication officers: A concurrent validity study.* Unpublished master's thesis, Radford University.

*Schuerger, J. M., Kochevar, K. F., & Reinwald, J. E. (1982). Male and female corrections officers: Personality and rated performance. *Psychological Reports, 51*(1), 223-228.

*Scogin, F., Schumacher, J., Gardner, J., & Chaplin, W. (1995). Predictive validity of psychological testing in law enforcement settings. *Professional Psychology: Research and Practice, 26*(1), 68-71.

*Serko, B. A. (1981). *Police selection: A predictive study.* Unpublished doctoral dissertation, Florida School of Professional Psychology.

*Shaffer, A. M. (1996). *Predictive and discriminative validity of various police officer selection criteria.* Unpublished master's thesis, University of California, Irvine.

*Shaver, D. P. (1980). *A descriptive study of police officers in selected towns of northwest Arkansas.* Unpublished doctoral dissertation, University of Arkansas.

*Shaw, J. H. (1986). Effectiveness of the MMPI in differentiating ideal from undesirable police officer applicants. In Reese, J. T. & Goldstein, H. A. (Eds). *Psychological services for law enforcement*, pp 91-95. Washington, D.C.: U.S. Government Printing Office.

*Sheppard, C., Bates, C., Fracchia, J., & Merlis, S. (1974). Psychological need structures of law enforcement officers. *Psychological Reports, 35,* 583-586.

*Shusman, E. J., Inwald, R. E., & Knatz, H. F. (1987). A cross-validation study of police recruit performance as predicted by the IPI and MMPI. *Journal of Police Science and Administration, 15*(2), 162-169.

*Shusman, E. J., Inwald, R. E., & Landa, B. (1984). Correction officer job performance as predicted by the IPI and MMPI: A validation and cross-validation study. *Criminal Justice and Behavior, 11*(3), 309-329.

*Simon, W. E., Wilde, V., & Cristal, R. M. (1973). Psychological needs of professional police personnel. *Psychological Reports, 33,* 313-314.

*Spielberger, C. D., Spaulding, H. C., Jolley, M. T., & Ward, J. C. (1979). Selection of effective law enforcement officers: The Florida police standards research project. In Charles D. Spielberger (Ed.). *Police Selection and Evaluation: Issues and Techniques.* New York: Praeger Publishers.

*Spielberger, C. D., Spaulding, H. C., Ward, J. C., & Vagg, P. R. (1981). *The Florida Police Standards Research Project: The Validation of a Psychological Test Battery for Selecting Law Enforcement Officers*. Tampa, Florida: University of South Florida.

*Sprenger, T. A. (1997). *Comparing personality profiles of law enforcement officers and criminals based on the MMPI and*

MMPI-2. Unpublished master's thesis, Emporia State University, Kansas.

*Sterrett, M. R. (1984). *The utility of the Bipolar Psychological Inventory for predicting tenure of law enforcement officers.* Unpublished doctoral dissertation, Claremont Graduate College.

*Stevenson, C. W. (1991). *A comparison of psychological characteristics of standout police officers and Oregon police academy trainees.* Unpublished doctoral dissertation, Oregon State University.

*Super, J. T. (1995). Psychological characteristics of successful SWAT/tactical response team personnel. *Journal of Police and Criminal Psychology, 11*(1), 60-63.

*Surrette, M. A., Aamodt, M. G., & Serafino, G. (1990). *Validity of the New Mexico Police Selection Battery.* Paper presented at the annual meeting of the Society for Police and Criminal Psychology, Albuquerque, NM.

Surrette, M. A., Ebert, J. M., Willis, M. A., & Smallidge, T. M. (2003). Personality of law enforcement officials: A comparison of law enforcement officials' personality profiles based on community size. *Public Personnel Management, 32*(2), 279-285.

*Sweda, M. G. (1988). *The Iowa law enforcement personnel study: Prediction of law enforcement job performance from biographical and personality variables.* Unpublished doctoral dissertation, University of Iowa.

*Talley, J. E., & Hinz, L. D. (1990). *Performance prediction of public safety and law enforcement personnel.* Springfield, IL: Charles C. Thomas.

*Tesauro, R. R. (1994). *The MMPI/MMPI-2 Immaturity Index as a predictor of police performance.* Unpublished doctoral dissertation, Tennessee State University.

*Tiburzi, M. J. (1996). *Use of pre-employment MMPI scores in predicting domestic violence perpetration in a large metropolitan police department.* Unpublished master's thesis, Loyola College.

*Tomini, B. A. (1995). *The person-job fit: Implications of selecting police personnel on the basis of job dimensions, aptitudes and personality traits.* Unpublished doctoral dissertation, University of Windsor.

*Topp, B. W., & Kardash, C. A. (1986). Personality, achievement, and attrition: Validation in a multiple-jurisdiction police academy. *Journal of Police Science and Administration, 14*(3), 234-241.

*Uno, E. A. (1979). *The prediction of job failure: A study of police officers using the MMPI.* Unpublished doctoral dissertation, California School of Professional Psychology-Berkeley.

*Varela, J. G., Scogin, F. R., & Vipperman, R. K. (1999). Development and preliminary validation of a semi-structured interview for the screening of law enforcement candidates. *Behavioral Science and the Law, 17*(4), 467-481.

*Vosburgh, B. V. (1987). *Police personality and performance: A concurrent validity study.* Unpublished doctoral dissertation, California School of Professional Psychology - Los Angeles.

*Ward, J. C. (1981). *The predictive validity of personality and demographic variables in the selection of law enforcement officers.* Unpublished doctoral dissertation, University of South Florida.

*Weiss, W. U., Davis, R., Rostow, C., & Kinsman, S. (2003). The MMPI-2 L scale as a tool in police selection. *Journal of Police and Criminal Psychology, 18*(1), 57-60.

*Weiss, W. I., Decoster, E., Davis, R., & Rostow, C. (2003, October). *The Personality Assessment Inventory as a selection device for law enforcement personnel.* Paper presented at the annual meeting of the Society for Police and Criminal Psychology, Corpus Christi, TX.

* Weiss, W. U., Serafino, G., Serafino, A., Wilson, W., & Knoll, S. (1998). Use of the MMPI-2 to predict the employment continuation and performance ratings of recently hired police officers. *Journal of Police and Criminal Psychology, 13*(1), 40-44.

*Weiss, W. U., Serafino, G., Serafino, A., Wilson, W., Sarsany, J., & Felton, J. (1999). Use of the MMPI-2 and the Inwald Personality Inventory to identify personality characteristics of dropouts from a state police academy. *Journal of Police and Criminal Psychology, 14*(1), 38-42.

*Wellman, R. J. (1982). *Accident proneness in police officers: Personality factors and problem drinking as predictors of injury claims of state troopers.* Unpublished doctoral dissertation, University of Connecticut.

*Wells, V. K. (1991). *The MMPI and CPI as predictors of police performance.* Unpublished doctoral dissertation, Saint Louis University.

*West, S. D. (1988). *The validity of the MMPI in the selection of police officers*. Unpublished master's thesis, University of North Texas.

*Wilson, A. (1980). *Reported accidental injuries in a metropolitan police department.* Unpublished doctoral dissertation, Boston University.

*Workowski, E. J., & Pallone, N. J. (1999). Previously unscored pre-service MMPI data in relation to police performance over a decade: A multivariate inquiry. *Journal of Offender Rehabilitation, 29*(3/4), 71-94.

*Wright, B. S. (1988). *Psychological evaluations as predictors of police recruit performance.* Unpublished doctoral dissertation, Florida State University.

*Wright, B. S., Doerner, W. G., Speir, J. C. (1990). Pre-employment psychological testing as a predictor of police performance during an FTO program. *American Journal of Police, 9*(4), 65-83.

*Zalen, G. W. (1967). *MMPI profile comparisons of the Michigan State Police compared for length of time-spent at command posts.* Unpublished doctoral dissertation, Central Michigan University.

*Study was used in the meta-analysis

Chapter 8
Vocational Interest Inventories

Vocational interest inventories are primarily used to help people choose careers that are best suited to their interests. Because interest inventories are easily faked, they are not often used in employee selection. In fact, for occupations in general, interest inventories have shown to have little validity ($r = .10$) in predicting employee performance (Hunter & Hunter, 1984). Though vocational interest inventories are not commonly used, there were several studies that investigated the validity of interest inventories in law enforcement selection so their validity for policing will be discussed in this chapter.

Interest Inventories Used in Law Enforcement Research

Aamodt Vocational Interest Inventory (AVIS)

The AVIS was designed to help working adults (rather than students) explore new careers following layoffs or voluntary career changes. The AVIS contains 130 questions yielding scores on 13 broad career dimensions: Clerical, customer service, science, analysis, sales, agriculture, transportation, trades, protection (e.g., police & fire), helping, leading, consumer economics, and creative. According to the test manual, the median test-retest reliability of the AVIS is .87 (range = .75 - .94). In the only use of the AVIS in law enforcement, Raynes (1997) found a correlation of -.09 between law enforcement interest and supervisor ratings of performance for 161 officers in three police departments.

Holland's Occupational Preference Inventory and Self-Directed Search

The Occupational Preference Inventory, which can be completed in 30 minutes, contains a list of 160 occupations for which test takers indicate their interest. The 160 questions produce scores on six main interest clusters (realistic, investigative, artistic, social, enterprising, and conventional) as well as scores on self-control, masculinity-femininity, status, infrequency, and acquiescence. Jobs in the realistic cluster involve those working with machines or one's hands. The job of police officer falls under this cluster. Investigative jobs involve science and computers, artistic jobs involve writing, music, and the arts; social jobs involve helping people (i.e., teacher), enterprising jobs involve sales and persuasion, and conventional jobs involve math and office work (i.e., accounting). In a study of 18 police officers from two small towns, Mass (1980) found significant correlations between scores on the investigative and artistic scales of the Vocational Preference Inventory and supervisor ratings of patrol performance.

The Self-Directed Search (SDS) yields scores on the six main interest clusters described above. The SDS has a median test-retest reliability of .82 and a median coefficient alpha of .92. Johnson and Hogan (1981) studied 50 police officers and 38 academy cadets and found that artistic interest scores were negatively related to academy performance (r = -.29) and positively related to complaints received (r = .34).

Strong Interest Inventory (SII)

The SII consists of 325 questions that yield scores on the 6 Holland themes, 23 basic interest groups, 207 specific occupations (including police), and four personal styles. The SII takes 20-60 minutes to complete and the median test-retest reliability is .91 for the basic interest scales and .92 for the occupational scales.

DuBois and Watson (1950) found negative correlations between scores on the police interest scale and academy grades (r

= - .09) and supervisor ratings of job performance (r = - .01), and a positive correlation with marksmanship (r = .12). Spaulding (1980) reported negative correlations between scores on the police interest scale and supervisor ratings of performance, and positive correlations with academy performance.

Kuder Occupational Interest Survey and Kuder Preference Record

The Kuder Preference Record was developed in 1948 and then replaced in 1971 by the Kuder Occupational Interest Survey. The inventory contains 100 questions, takes 30-45 minutes to complete, and yields scores on 119 occupational groups (including police) and 10 occupational areas (outdoor, mechanical, computational, scientific, persuasive, artistic, literary, musical, and clerical). In research using the Kuder Occupational Interest Survey, Azen, Snibbe, & Montgomery (1973) found a correlation of .24 between the Mechanical Interest scale and supervisor ratings of performance. In research using the Kuder Preference Record, Sterne (1960) found that scores on preference with dealing with ideas and preference for directing others were correlated with supervisor ratings on the same dimensions, and Serko (1981) found a negative correlation between a preference for working with ideas and supervisor ratings of performance.

Meta-Analysis Results

Eight studies were found that investigated the validity of vocational interest. These eight studies yielded 58 separate correlations. As shown in Table 8.1, with the exception of the conventional interest area, vocational interest does not appear to be positively related to supervisor ratings of job performance. From Table 8.1, three findings stand out. One, conventional interests were significantly related to job performance (r = .14, ρ = .25). Two, law enforcement interest was negatively related, although not significantly so, to police performance (r = - .03, ρ = - .06). And three, artistic interests were negatively related to police performance. Although interest inventories are easily faked, the positive correlation with conventional interests and the negative correlation with artistic interests ($r = -.12$, $\rho = -.22$) suggest that further research on the validity and potential use of interest inventories might be a good idea.

Table 8.1: Meta-analysis results for the validity of vocational interest in predicting supervisor ratings of performance

Interest Scale	K	N	r	95% Confidence Interval		ρ	90% Credibility Interval		Var	Q_w
				Lower	Upper		Lower	Upper		
All Interest Scales	58	5,336	- .04	- .07	.00	- .06	- .25	.13	71%	82.12*
Police interest	8	825	- .03	- .10	.03	- .06	- .06	- .06	100%	2.82
Realistic	5	332	- .02	- .13	.09	- .03	- .03	- .03	100%	3.48
Investigative	5	332	- .01	- .12	.10	- .02	- .02	- .02	100%	5.01
Artistic	5	332	- .12	- .23	- .01	- .22	- .38	- .05	84%	5.97
Social	5	332	.03	- .08	.13	.04	.04	.04	100%	2.38
Enterprising	5	332	- .03	- .14	.08	- .05	- .05	- .05	100%	2.72
Conventional	5	332	.14	.03	.24	.25	.25	.25	100%	0.70

K=number of studies, N=sample size, r = mean correlation, ρ = mean correlation corrected for range restriction, criterion unreliability, and predictor reliability, VAR = percentage of variance explained by sampling error and study artifacts, Q_w = the within group heterogeneity

Chapter References

*Azen, S. P., Snibbe, H. M., & Montgomery, H. R. (1973). A longitudinal predictive study of success and performance of law enforcement officers. *Journal of Applied Psychology, 57*(2), 190-192.

*DuBois, P. H., & Watson, R. I. (1950). A longitudinal predictive study of success and performance of law enforcement officers. *Journal of Applied Psychology, 34*(1), 90-95.

Hunter, J. E., & Hunter, R. F. (1984). Validity and utility of alternative predictors of job performance. *Psychological Bulletin, 96(*1), 72-98.

*Johnson, J. A., & Hogan, R. (1981). Vocational interests, personality, and effective police performance. *Personnel Psychology, 34*(1), 49-53.

*Mass, G. (1980). *Using judgment and personality measures to predict effectiveness in policework: An exploratory validation study.* Unpublished doctoral dissertation, Ohio State University.

*Raynes, B. J. (1997).). *Predicting difficult employees: the relationship between vocational interest, self-esteem, and problem communication styles.* Unpublished master's thesis, Radford University.

*Serko, B. A. (1991). *Police selection: A predictive study.* Unpublished doctoral dissertation, Florida School of Professional Psychology.

Spaulding, H. C. (1980). Predicting police officer performance: The development of screening and selection procedures based on criterion-related validity.* Unpublished master's thesis, University of South Florida.

* Sterne, D. M. (1960). Use of the Kuder Preference Record, Personal, with police officers. *Journal of Applied Psychology, 44*(5), 323-324.

*Indicates study was included in the chapter meta-analysis

Chapter 9
Assessment Centers

An assessment center is a selection method in which an applicant is assessed on several exercises, at least one of which must be a job simulation. Applicants' performance on these exercises is evaluated by multiple, trained assessors. Common exercises in assessment centers include leaderless group discussions, role plays, work samples, in-baskets, and interviews. Because assessment centers typically require that applicants already have job related experience, they are usually used for promotion purposes rather than entry-level selection. This is especially true in law enforcement where successful applicants are expected to have the ability to perform as police officers and are then taught the necessary knowledge and skills during the academy and field training. Because assessment centers are seldom used for entry-level selection, there are few studies related to assessment center validity and entry-level law enforcement performance. This chapter presents a meta-analysis of those few studies.

Meta-Analysis Results

Only six studies were found investigating the validity of assessment centers in a law enforcement setting. Because only six studies were found, I validated the comprehensiveness of my literature review by reviewing the bibliographies of two meta-analyses on assessment center validity in all job settings (Arthur, Day, McNelly, & Edens, 2003; Gaugler, Rosenthal, Thornton, & Bentson, 1987). No new law enforcement related studies were found in either bibliography.

Two of the six studies found for this meta-analysis (Foster, 1995; Ross, 1980) looked at assessment centers for promotional purposes, so these studies were separated from the four using assessment centers for entry-level hiring. As can be seen in Table 9.1, assessment center ratings significantly predict academy ratings, entry-level job performance, and performance of lieutenants and captains. The validity of assessment center scores ($r = .33$, $\rho = .38$) when used for promotion purposes is similar to the validity of overall assessment center ratings obtained by Arthur, et al. (2003; $r = .28$, $\rho = .38$) and by Gaugler et al. (1987; $r = .25$, $\rho = .36$) in their meta-analyses of the validity of assessment center scores across all types of jobs. Because of the small number of studies in the current meta-analysis, much more research is needed in this area, especially research that investigates the validity of scores on specific assessment center dimensions rather than overall assessment center scores (Arthur et al., 2003). Furthermore, because the typical assessment center costs $1,750 and can take as long as two days to complete (Spychalski, Quinones, Gaugler, & Pohley, 1997), it is important to determine if the validity of assessment centers justifies their costs.

Table 9.1: Meta-analysis results for the validity of assessment centers

Criteria	K	N	r	95% Confidence Interval		ρ	90% Credibility Interval		Var	Q_w
				Lower	Upper		Lower	Upper		
Academy Grades	2	784	.22	.16	.29	.37	.37	.37	100%	0.74
Supervisor Ratings										
Entry-level	4	785	.17	.11	.24	.31	.31	.31	100%	1.20
Promotion	2	388	.33	.24	.42	.57	.57	.57	100%	0.40

K=number of studies, N=sample size, r = mean correlation, ρ = mean correlation corrected for range restriction, criterion unreliability, and predictor reliability, VAR = percentage of variance explained by sampling error and study artifacts, Q_w = the within group heterogeneity

Chapter References

Arthur, W., Day, E. A., McNelly, T. L., & Edens, P. S. (2003). A meta-analysis of the criterion-related validity of assessment center dimensions. *Personnel Psychology, 56*(1), 125-154.

*Bromley, M. (1996). Evaluating the use of the assessment center process for entry-level police officer selection in a medium sized police agency. *Journal of Police and Criminal Psychology, 10*(4), 33-40.

*Dayan, K., Kasten, R., & Fox, S. (2002). Entry-level police candidate assessment center: An efficient tool or a hammer to kill a fly. *Personnel Psychology, 55*(4), 827-849.

*Foster, M. R. (1995). *The use of biographical information to determine skill levels as measured in an assessment center*. Unpublished doctoral dissertation, University of Georgia.

Gaugler, B. B., Rosenthal, D. B., Thornton, G. C., & Bentson, C. (1987). Meta-analysis of assessment center validity. *Journal of Applied Psychology, 72*, 493-511.

*Holland, A. M. (1980). *Comparative analysis of selected predictors of police officer job performance.* Unpublished doctoral dissertation, Georgia State University.

*Pynes, J., & Bernardin, H. J. (1989). Predictive validity of an entry-level police officer assessment center. *Journal of Applied Psychology, 74*(5), 831-833.

*Ross, J. D. (1980). Determination of the predictive validity of the assessment center approach to selecting police managers. *Journal of Criminal Justice, 8*(1), 89-96.

Spychalski, A. C., Quinones, M. A., Gaugler, B. B., & Pohley, K. (1997). A survey of assessment center practices in organizations in the United States. *Personnel Psychology, 50*(1), 71-90.

*Study was used in meta-analysis

Chapter 10
Interviews

There are few jobs in which applicants don't go through some type of interview, and law enforcement positions are no exception. Interviews can be differentiated from one another on the basis of their structure and the number of people conducting the interview.

Structure

Structured interviews base their questions on a job analysis so that all questions are job related, every applicant is asked the same questions, and there is a standard scoring procedure for each question (Aamodt, 2004). Ideally, interviewers participating in structured interviews have received extensive training on how to interact with applicants and how to score answers to interview questions. Unstructured interviews are missing at least one of the above three components. Previous meta-analysis results clearly indicate that structured interviews are more valid than unstructured interviews (Huffcutt & Arthur, 1994; McDaniel, Whetzel, Schmidt, & Maurer, 1994).

Number of Interviewers

Interviews usually come in one of three formats on the basis of the number of interviewers. Individual interviews involve a single person interviewing a candidate and making a recommendation. Sequential interviews involve multiple interviewers, each of whom interviews candidates separately. Panel interviews involve multiple interviewers who interview the applicant in a group setting. All interviews found for this meta-analysis were panel interviews.

Meta-Analysis Results

Eight studies were located investigating the validity of interview scores with either academy performance or supervisor ratings of job performance. As shown in Table 10.1, interview scores were significantly related to both academy and supervisor ratings of job performance. Though both coefficients are statistically significant, they are relatively small. Clearly, more research is needed on this topic. As mentioned previously, an abundance of research has demonstrated that structured interviews are one of the most valid predictors of employee performance. Such is probably the case with law enforcement. However, in the interview studies reviewed here it was unclear to what extent the interviews were structured.

In addition to panel interviews conducted by police departments or civil service commissions, clinical interviews are often conducted by psychologists to determine potential pathology. As shown in Table 10.2, not only did these types of interviews not significantly predict performance, when they were used with a test battery, they may reduce the validity of the test battery. Though more research is needed, it appears that clinical interviews are not useful in selecting law enforcement personnel.

Table 10.1: Meta-analysis results for the validity of interviews

Criteria	K	N	r	95% Confidence Interval		ρ	90% Credibility Interval		Var	Q_w
				Lower	Upper		Lower	Upper		
Academy Grades	4	554	.12	.04	.20	.27	.27	.27	100%	2.33
Supervisor Ratings	8	1,053	.09	.03	.15	.19	.19	.19	100%	1.80

K=number of studies, N=sample size, r = mean correlation, ρ = mean correlation corrected for range restriction, criterion unreliability, and predictor reliability, VAR = percentage of variance explained by sampling error and study artifacts, Q_w = the within group heterogeneity

Table 10.2: Meta-analysis results for the validity of interviews conducted by psychologists

Criteria	K	N	r	95% Confidence Interval		ρ	90% Credibility Interval		Var	Q_w
				Lower	Upper		Lower	Upper		
Clinical interview	4	486	.05	- .04	.14	.10	.10	.10	100%	4.00
Test battery	30	6,668	.18	.13	.23	.35	.03	.68	42%	71.18*
Interview + test battery	6	741	.12	.06	.19	.24	.24	.24	100%	4.27

K=number of studies, N=sample size, r = mean correlation, ρ = mean correlation corrected for range restriction, criterion unreliability, and predictor reliability, VAR = percentage of variance explained by sampling error and study artifacts, Q_w = the within group heterogeneity

Chapter References

Aamodt, M. G. (2004). *Applied industrial/organizational psychology* (4th ed). Belmont, CA: Wadsworth Publishing.

*Cave, S. B., & Westfried, E. (2002). Linkage between pre-employment evaluations, academy performance, and first year job performance ratings with a state police agency. Paper presented at the annual meeting of the Society for Police and Criminal Psychology, Orlando, FL.

*Davidson, N. B. (1975). *The predictive validity of a police officer selection program.* Unpublished master's thesis, Portland State University.

*Flynn, J. T., & Peterson, M. (1972). The use of regression analysis in police patrolman selection. *Journal of Criminal Law, 63*(4), 564-569.

*Giannoni, R. J. (1979). *Personnel selection procedures and their relationship with academy training and field performance of state traffic officers.* Unpublished master's thesis, California State University, Sacramento.

*Hess, L. R. (1972). *Police entry tests and their predictability of score in police academy and subsequent job performance.* Unpublished doctoral dissertation, Marquette University.

Huffcutt, A. I., & Arthur, W. (1994). Hunter and Hunter (1984) revisited: Interview validity for entry level jobs. *Journal of Applied Psychology, 79*(2), 184-190.

*Kayode, O. (1973). *Predicting performance on the basis of social background characteristics: the case of the Philadelphia Police Department.* Unpublished doctoral dissertation, University of Pennsylvania.

*Knights, R. M. (1976). *The relationship between the selection process and on-the-job performance of Albuquerque police officers.* Unpublished doctoral dissertation, University of New Mexico.

*Landy, F. J. (1976). The validity of the interview in police officer selection. *Journal of Applied Psychology, 61*(2), 193-198.

McDaniel, M. A., Whetzel, D. L., Schmidt, F. L., & Maurer, S. D. (1994). The validity of employment interviews: A comprehensive review and meta-analysis. *Journal of Applied Psychology, 79*, 599-616.

*Study used in meta-analysis

Chapter 11
Physical Agility

Most law enforcement agencies require applicants to pass a physical agility test either prior to hire or at the completion of the academy. These physical agility tests usually come in one of two formats. In the first format, applicants perform a variety of exercises related to stamina and strength. Such exercises often include sprints, push-ups, and sit-ups. In the second format, applicants perform a job-related simulation involving physical agility. Such a simulation might involve getting out of a car, running a short distance, leaping over an obstacle, climbing a fence, going through a window, and dry-firing a pistol at the end of the obstacle course. This chapter will explore the validity of these physical agility tests.

Meta-Analysis Results

Though many studies have examined the *content* validity of physical ability tests (e.g., Anderson, Plecas, & Segger, 2001), only four studies were found that investigated the *criterion* validity of physical agility test scores. As shown in Table 11.1, on the basis of these four studies, there appears to be no relationship between overall physical agility scores and supervisor ratings of overall job performance. However, two studies (Arvey, Landon, Nutting, & Maxwell, 1992; Jeanneret, Moore, Blakley, Koelzer, & Menkes, 1991) listed the validities for the individual components of their physical agility tests, and these results tell a more positive story. As can be seen in Table 11.1, some of the individual physical agility components are significant predictors of performance. Furthermore, job-related simulations seem to be better predictors than direct tests of specific physical abilities (e.g., grip strength, sit-ups).

Data from Arvey, et al. (1992) and Jeanneret et al. (1991) suggest that physical agility tests are highly related to supervisor ratings of physical ability. Median correlations between physical ability tests and supervisor ratings of physical fitness of .35 and .38 were found by Jeanneret et al. and Arvey et al. respectively.

Because physical agility tests are commonly used and have excellent content and face validity, more research is needed to determine their criterion validity.

Table 11.1: Meta-analysis results for the validity of physical agility tests in predicting supervisor ratings

Criterion	K	N	r	95% Confidence Interval		ρ	90% Credibility Interval		Var	Q_w
				Lower	Upper		Lower	Upper		
Overall fitness score	4	547	-.02	-.16	.11	-.04	-.30	.23	39%	10.16*
Individual tests	23	3,464	.09	.05	.14	.16	-.02	.33	64%	36.07*
Running	3	408	.16	.07	.26	.26	.26	.26	100%	1.48
Sit-ups	2	293	.07	-.05	.18	.11	.11	.11	100%	0.96
Strength	14	2,177	.09	.04	.15	.15	-.01	.31	67%	21.26
Low body fact	1	115	.33							
Category of Test										
Physical ability	16	2,407	.08	.06	.14	-.06	-.06	.34	57%	28.26*
Simulation	7	1,057	.12	.06	.20	.20	.20	.20	100%	7.00

K=number of studies, N=sample size, r = mean correlation, ρ = mean correlation corrected for range restriction, criterion unreliability, and predictor reliability, VAR = percentage of variance explained by sampling error and study artifacts, Q_w = the within group heterogeneity

Chapter References

Anderson, G. S., Plecas, D., & Segger, T. (2001). Police officer physical ability testing: Revalidating a selection criterion. *Policing: An International Journal of Police Strategies & Management, 24*(1), 8-31.

*Arvey, R. D., Landon, T. E., Nutting, S. M., & Maxwell, S. E. (1992). Development of physical ability tests for police officers: A construct validation approach. *Journal of Applied Psychology, 77*(6), 996-1009.

*Bertram, F. D. (1975). *The prediction of police academy performance and on-the-job performance from police recruit screening measures.* Unpublished doctoral dissertation, Marquette University.

*Jeanneret, P. R., Moore, J. R., Blakley, B. R., Koelzer, S. L., & Menkes, O. (1991). *Development and validation of trooper physical ability and cognitive ability tests: Final report submitted to the Texas Department of Public Safety.* Houston, TX: Jeanneret & Associates.

*Wexler, N., & Sullivan, S. M. (1982). *Concurrent validation of a prototype selection test for entry-level police officer.* Trenton, NJ: New Jersey Department of Civil Service, Division of Examinations.

*Study was included in the meta-analysis

Chapter 12
Correlations Among Performance Criteria

When conducting performance appraisals, it is important for supervisors to understand how various performance measures are related to one another. That is, is an officer with an above average number of citizen complaints a bad officer or is the number of complaints simply an artifact of making more arrests and issuing more tickets? This chapter reports the results of a meta-analysis investigating the relationships among academy grades, probationary performance, supervisor ratings, commendations, problems (e.g., complaints, disciplinary actions, suspensions), absenteeism, accidents and injuries, and use of force.

Meta-Analysis Findings

Forty-seven studies were located that listed at least one correlation between two different performance criteria. As shown in Table 12.1, performance in the academy is significantly related to on-the-job performance. Officers performing well in the academy can be expected to receive higher performance ratings ($r = .22$, $\rho = .38$), have fewer disciplinary problems ($r = -.13$, $\rho = -.22$), make more arrests ($r = .24$, $\rho = .40$), and receive more commendations ($r = .07$, $\rho = .10$) than their peers who did not perform as well in the academy. These are important findings because they justify the use of academy grades as a criterion when validating selection tests.

The correlations between supervisor ratings of job performance and commendations, problems, and activity are interesting. Though the correlations with supervisor ratings are in general statistically significant, they are not as high as one would imagine. Thus, it would appear that the various performance measures are relatively independent.

Another interesting finding is the correlations between discipline problems and commendations and activity. The positive correlation between discipline problems and activity (r = .17, ρ = .31) indicates that officers making more arrests and issuing more citations receive a greater number of complaints than do their less active counterparts. The low positive correlation between discipline problems and commendations (r = .05, ρ = .07) suggests that the two are not inversely related. That is, officers who receive a lot of commendations are not necessarily getting few complaints.

Table 12.1: Meta-analysis results of the relationship among performance measures

Criterion	K	N	r	95% Confidence Interval		ρ	90% Credibility Interval		Var	Q_w
				Lower	Upper		Lower	Upper		
Academy Grades										
FTO Performance	7	1,911	.25	.17	.33	.44	.20	.48	80%	8.80
Supervisor ratings	19	5,012	.22	.18	.26	.38	.38	.38	100%	17.18
Discipline problems	10	5,553	-.13	-.18	-.08	-.22	-.33	.00	45%	22.44*
Commendations	6	4,217	.07	.01	.14	.10	-.09	.24	25%	23.97*
Activity	3	356	.24	.14	.34	.40	.40	.40	100%	1.57
Absenteeism	1	1,608	-.13							
Supervisor Ratings										
FTO Performance	2	165	.26							
Discipline problems	17	6,568	-.10	-.16	-.04	-.16	-.48	.18	21%	93.24*
Commendations	12	4,023	.21	.15	.26	.30	.00	.45	43%	32.47*
Activity	8	1,046	.14	-.05	.34	.27	-.55	.99	12%	65.83*
Absenteeism	9	2,205	-.05	-.02	.12	-.09	-.31	.14	38%	23.54*
Peer-ratings	2	150	.46							
Self-ratings	2	73	.37							

K=number of studies, N=sample size, r = mean correlation, ρ = mean correlation corrected for range restriction, criterion unreliability, and predictor reliability, VAR = percentage of variance explained by sampling error and study artifacts, Q_w = the within group heterogeneity

Table 12.1: Meta-analysis results of the relationship among performance measures (continued)

Criterion	K	N	r	95% Confidence Interval		ρ	90% Credibility Interval		Var	Q_w
				Lower	Upper		Lower	Upper		
Discipline Problems										
Commendations	35	12,685	.05	.01	.08	.07	- .19	.33	22%	157.36*
Absenteeism	22	8,579	.13	.06	.20	.20	- .18	.59	13%	165.70*
Activity	7	1,210	.17	.10	.24	.31	.17	.45	78%	8.94
Commendations										
Absenteeism	10	2,494	- .02	- .06	.02	- .02	- .02	- .02	100%	7.87
Activity	1	31	.23							
Peer ratings	1	52	.19							
Self-ratings	1	52	- .13							
Activity										
Absenteeism	3	385	.01	- .20	.21	.01	- .39	.41	24%	12.53*

K=number of studies, N=sample size, r = mean correlation, ρ = mean correlation corrected for range restriction, criterion unreliability, and predictor reliability, VAR = percentage of variance explained by sampling error and study artifacts, Q_w = the within group heterogeneity

Chapter References

*Abbatiello, A. A. (1969). *A study of police candidate selection.* Paper presented at the 77th Annual Convention of the American Psychological Association, Washington, D. C.

*Baehr, M. E., Furcon, J. E., & Froemel, E. C. (1968). *Psychological assessment of patrolman qualifications in relation to field performance.* Washington, D.C.: Law Enforcement Assistance Administration.

*Beutler, L. Storm, A., Kirkish, P., Scogin, F., & Gaines, J. A. (1985). Parameters in the prediction of police officer performance. *Professional Psychology: Research and Practice, 16*(2), 324-335.

*Buttolph, S. E. (1999). *Effect of college education on police behavior: Analysis of complaints and commendations.* Unpublished master's thesis, East Tennessee State University.

*Campa, E. E. (1993). *The relationship of reading comprehension and educational achievement levels to academy and field training performance of police cadets.* Unpublished doctoral dissertation, Texas A&M University.

*Champion, D. H. (1994). *A study of the relationship between critical thinking levels and job performance of police officers in a medium size police department in North Carolina.* Unpublished doctoral dissertation, North Carolina State University.

*Clopton, W. (1971). *Comparison of ratings and field performance data in validating predictions of patrolman performance: A five-year follow-up study.* Unpublished master's thesis, University of Cincinnati.

*Cohen, B. & Chaiken, J. M. (1973). *Police Background Characteristics and Performance.* Lexington, MA: Lexington Books.

*Copley, W. H. (1987). *Using education, academy, and field training scores to predict success in a Colorado police department.* Unpublished doctoral dissertation, Colorado State University.

*Cortina, J. M., Doherty, M. L, Schmitt, N., Kaufman, G., & Smith, R. G. (1992). The "Big Five" personality factors in the IPI and MMPI: Predictors of police performance. *Personnel Psychology, 45,* 119-140.

*Culley, J. A. (1987). *Height standards and policing: Rationale or rationalization?* Unpublished doctoral dissertation, SUNY - Albany.

*Daley, R. E. (1978). *The relationship of personality variables to suitability for police work.* Unpublished doctoral dissertation, Florida Institute of Technology.

*Dayan, K., Kasten, R., & Fox, S. (2002). Entry-level police candidate assessment center: An efficient tool or a hammer to kill a fly? *Personnel Psychology, 55*(4), 827-849.

*Fabricatore, J., Azen, S., Schoentgen, S., & Snibbe, H. (1978). Predicting performance of police officers using the Sixteen Personality Factor Questionnaire. *American Journal of Community Psychology, 6*(1), 63-69.

*Feehan, R. L. (1977). *An investigation of police performance utilizing mental ability selection scores, police academy training scores, and supervisory ratings of the job performance of patrol officers.* Unpublished doctoral dissertation, Georgia Institute of Technology.

*Ford, J. K., & Kraiger, K. (1993). Police officer selection validation project: The multijurisdictional police officer examination. *Journal of Business and Psychology, 7*(4), 421-429.

*Garber, C. R. (1983). *Correlation studies using entry scores, training test results, and subsequent job performance ratings of students of the security police academy, Lackland AFB, Texas.* Unpublished doctoral dissertation, Brigham Young University.

*Giannoni, R. J. (1979). *Personnel selection procedures and their relationship with academy training and field performance of state traffic officers.* Unpublished master's thesis, California State University, Sacramento.

*Hankey, R. O. (1968). *Personality correlates in a role of authority: The police*. Unpublished doctoral dissertation, University of Southern California.

*Hargrave, G. E., Norborg. J. M., & Oldenburg, L. (1986). Differences in entry level test and criterion data for male and female police officers. In Reese, J. T. & Goldstein, H. A. (Eds). *Psychological services for law enforcement*, pp 35-42. Washington, D.C.: U.S. Government Printing Office.

*Hausknecht, J. P., Trevor, C. O., & Farr, J. L. (2002). Retaking ability tests in a selection setting: Implications for practice effects, training performance, and turnover. *Journal of Applied Psychology, 87*(2), 243-254.

*Holland, A. M. (1980). *Comparative analysis of selected predictors of police officer job performance*. Unpublished doctoral dissertation, Georgia State University.

*Hooper, M. K. (1988). *Relationship of college education to police officer job performance*. Unpublished doctoral dissertation, Claremont Graduate School.

*Jeanneret, P. R., Moore, J. R., Blakley, B. R., Koelzer, S. L., & Menkes, O. (1991). *Development and validation of trooper physical ability and cognitive ability tests: Final report submitted to the Texas Department of Public Safety*. Houston, TX: Jeanneret & Associates.

*Johnson, J. A., & Hogan, R. (1981). Vocational interests, personality, and effective police performance. *Personnel Psychology, 34*(1), 49-53.

*Kedia, P. R. (1985). *Assessing the effect of college education on police job performance*. Unpublished doctoral dissertation, University of Southern Mississippi.

*Kleiman, L. S. (1978). *Ability and personality factors moderating the relationships of police academy training performance with measures of selection and job performance*. Unpublished doctoral dissertation, University of Tennessee, Knoxville.

*Kleiman, L. S., & Gordon, M. E. (1986). An examination of the relationship between police training academy performance and job performance. *Journal of Police Science and Administration, 14*(4), 293-299.

*Leitner, D. W., & Sedlacek, W. E. (1976). Characteristics of successful campus police officers. *Journal of College Student Personnel, July*, 304-308.

*Mandel, K. (1970). *The predictive validity of on-the-job performance of policemen from recruitment selection information.* Unpublished doctoral dissertation, University of Utah.

*Mass, G. (1980). *Using judgment and personality measures to predict effectiveness in police work: An exploratory validation study.* Unpublished doctoral dissertation, Ohio State University.

*McEuen, O. L. (1981). *Assessment of some personality traits that show a relationship to academy grades, being dismissed from the department, and work evaluation ratings for police officers in Atlanta, Georgia.* Unpublished doctoral dissertation, The Fielding Institute.

*Murrell, D. B. (1982). The influence of education on police work performance. Unpublished doctoral dissertation, Florida State University.

*Palmatier, J. J. (1996). *The big-five factors and hostility in the MMPI and IPI: Predictors of Michigan State Trooper job performance.* Unpublished doctoral dissertation, Michigan State University.

*Parviainen, W. J. (1979). *The relationship between recruit school evaluations and future job performance in predicting job success for Michigan sate police troopers.* Unpublished master's thesis, Michigan State University.

*Peterson, D. S. (2001). *The relationship between educational attainment and police performance.* Unpublished doctoral dissertation, Illinois State University.

*Powers, W. P. (1996). *An evaluation of the predictive validity of the MMPI s it relates to identifying police officers prone to engage in the use of excessive force.* Unpublished doctoral dissertation, Adler School of Professional Psychology.

*Pynes, J. (1988). *The predictive validity of an assessment center for the selection of entry-level law enforcement officers.* Unpublished doctoral dissertation, Florida Atlantic University.

*Schelling, M. J. (1993). *The use of the Law Enforcement Selection Inventory in the selection of communication officers: A concurrent validity study.* Unpublished master's thesis, Radford University.

*Schumacher, J. E., Scogin, F., Howland, K., & McGee, J. (1992). The relation of peer assessment to future law enforcement performance. *Criminal Justice and Behavior, 19*(3), 286-293.

*Shaver, D. P. (1980). *A descriptive study of police officers in selected towns of northwest Arkansas.* Unpublished doctoral dissertation, University of Arkansas.

*Spurlin, O., & Swander, C. (1987). *Validity and fairness of the police officer written exam: Research finding.* Seattle, WA: Public Safety Civil Service Commission.

*Staff, T. G. (1992). *The utility of biographical data in predicting job performance: Implications for the selection of police officers.* Unpublished doctoral dissertation, University of Toledo.

*Tompkins, L. P. (1986). *An evaluation of police academy training upon selected recruits and its relationship to job performance.* Unpublished master's thesis, Rollins College, FL.

*Truxillo, D. M., Bennett, S. R., & Collins, M. L. (1998). College education and police job performance: A ten-year study. *Public Personnel Management, 27*(2), 269-280.

*Varela, J. G., Scogin, F. R., & Vipperman, R. K. (1999). Development and preliminary validation of a semi-structured interview for the screening of law enforcement candidates. *Behavioral Science and the Law, 17*(4), 467-481.

* Workowski, E. J., & Pallone, N. J. (1999). Previously unscored pre-service MMPI data in relation to police performance

over a decade: A multivariate inquiry. *Journal of Offender Rehabilitation, 29*(3/4), 71-94.

*indicates that the reference was used in the meta-analysis

Chapter 13 Correlations among Predictors of Performance

The studies summarized at the end of this book include a wide variety of methods used to predict law enforcement performance. These methods included background variables, education, cognitive ability, interviews, psychological suitability ratings, physical agility, personality inventories, and interest inventories. The meta-analysis for this chapter explores the relationships among these predictors.

Meta-Analysis Results

Only 21 studies provided information on the correlations among some of the predictors (selection methods). As can been seen in Table 13.1, the various predictors are not highly correlated, indicating that education, cognitive ability, panel interviews, clinician ratings, and physical agility tests appear to tap separate constructs. The relatively high correlations between cognitive ability and clinician ratings of psychological suitability and between clinician ratings of suitability and interview scores need to be interpreted with caution as one large study had an unusually high correlation ($r = .80$) between the variables. For most of the analyses, there was tremendous variability in correlations among the studies, making interpretation even more difficult. Because of the small number of studies, a search for moderators that would explain some of this variability was not practical. Due to the small number of studies and the high variability, this is clearly an area needing further research.

Table 13.1: Meta-analysis results of the relationships among predictors

Criterion	K	N	*r*	95% Confidence Interval Lower	95% Confidence Interval Upper	ρ	90% Credibility Interval Lower	90% Credibility Interval Upper	Var	Q_w
Cognitive Ability										
Education	12	4,776	.29	.23	.36	.42	.25	.60	55%	21.63*
Interview scores	9	1,632	.29	.10	.48	.40	- .29	.99	19%	47.88*
Clinician ratings	5	886	.42	.09	.74	.68	- .25	.99	13%	38.58*
Physical ability	4	873	- .15	- .41	.10	- .23	- .83	.38	8%	47.36*
Military service	2	3,122	.05	.02	.09	.07	.07	.07	100%	0.39
Education										
Military service	6	4,146	- .04	- .11	.03	- .06	- .24	.13	18%	33.88*
Physical ability	2	232	.07	- .06	.20	.10	.10	.10	100%	0.12
Interview scores	1	437	- .14							
Interview Scores										
Clinician ratings	6	1,107	.44	.17	.71	.78	- .06	.99	28%	21.39*
Physical ability	3	862	.13	- .02	.27	.21	- .08	.49	33%	9.05*
Clinician Ratings										
Physical ability	2	641	.29	.15	.43	.44	.32	.56	77%	2.60

K=number of studies, N=sample size, r = mean correlation, ρ = mean correlation corrected for range restriction, criterion unreliability, and predictor reliability, VAR = percentage of variance explained by sampling error and study artifacts, Q_w = the within group heterogeneity

Chapter References

*Aamodt, M. G., & Kimbrough, W. W. (1990). *Development of a police selection battery: A ten-year follow-up.* Paper presented at the annual meeting of the Society for Police and Criminal Psychology, Albuquerque, New Mexico.

*Barbas, C. (1992). *A study to predict the performance of cadets in a police academy using a modified CLOZE reading test, a civil service aptitude test, and educational level.* Unpublished doctoral dissertation, Boston University.

*Cascio, W. F. (1977). Formal education and police officer performance. *Journal of Police Science and Administration, 5*(1), 89-96.

*Champion, D. H. (1994). *A study of the relationship between critical thinking levels and job performance of police officers in a medium size police department in North Carolina.* Unpublished doctoral dissertation, North Carolina State University.

*Dailey, J. D. (2002). *An investigation of police officer background and performance: An analytical study of the effect of age, time in service, prior military service, and educational level on commendations.* Unpublished doctoral dissertation, Sam Houston State University.

*Davidson, N. B. (1975). *The predictive validity of a police officer selection program.* Unpublished master's thesis, Portland State University.

*Davis, R. D., & Rostow, C. D. (2003). Relationship between cognitive ability and background variables and disciplinary problems in law enforcement. *Applied H.R.M. Research, 8*(2), 77-80.

*Flynn, J. T., & Peterson, M. (1972). The use of regression analysis in police patrolman selection. *Journal of Criminal Law, 63*(4), 564-569.

*Gruber, G. (1986). The police applicant test: A predictive validity study. *Journal of Police Science and Administration, 14*(2), 121-129.

*Hausknecht, J. P., Trevor, C. O., & Farr, J. L. (2002). Retaking ability tests in a selection setting: Implications for practice effects, training performance, and turnover. *Journal of Applied Psychology, 87*(2), 243-254.

*Ho, T. (2001). The interrelationships of psychological testing, psychologists' recommendations, and police departments' recruitment decisions. *Police Quarterly, 4*(3), 318-342.

*Jayaraj, E. A. S. (1984). *A predictive study of police officers selection.* Unpublished master's thesis, Southern Connecticut State University.

*Kedia, P. R. (1985). *Assessing the effect of college education on police job performance.* Unpublished doctoral dissertation, University of Southern Mississippi.

*Knights, R. M. (1976). *The relationship between the selection process and on-the-job performance of Albuquerque police officers.* Unpublished doctoral dissertation, University of New Mexico.

*Mullineaux, J. E. (1965). An evaluation of the predictors used to select patrolmen. *Public Personnel Review, 16*, 84-86.

*Patterson, G. T. (2002). Predicting the effects of military experience on stressful occupational events in police officers. *Policing: An International Journal of Police Strategies & Management, 25*(3), 602-618.

*Rose, J. E. (1995). *Consolidation of law enforcement basic training academies: An evaluation of pilot projects.* Unpublished doctoral dissertation, Northern Arizona University.

*Scarfo, S. J. (2002). Relationship between police academy performance and cadet level of education and cognitive ability. *Applied H.R.M. Research, 7*(1), 24.

*Truxillo, D. M., Bennett, S. R., & Collins, M. L. (1998). College education and police job performance: A ten-year study. *Public Personnel Management, 27*(2), 269-280.

*Waugh, L. (1996). *Police officer recruit selection: Predictors of academy performance.* Queensland, Australia: Queensland Police Academy.

*Wexler, N., & Sullivan, S. M. (1982). *Concurrent validation of a prototype selection test for entry-level police officer.* Trenton, NJ: New Jersey Department of Civil Service, Division of Examinations.

*Study was used in the meta-analysis

Chapter 14
Sex, Race, Age, and Tenure

This chapter reports the results of a meta-analysis investigating the relationships between the criterion variables discussed in this book and sex, race, age, and years on the force. The initial thought was to also include correlations with the predictors but not enough information was provided in the studies to make this a meaningful analysis.

Meta-Analysis Results

To make it easier to interpret the results, sex was coded as a 0 for men and a 1 for women and race was coded as 0 for whites and 1 for minorities for all studies. Thus, negative correlations for sex and race indicate that women and minorities score lower on something. For example, a negative correlation between sex and academy performance would indicate that as a group, women performed worse in the academy than men.

As shown in Table 14.1, women and minority officers had lower academy averages and lower performance ratings than men and white officers. The finding for race is consistent with a meta-analysis by Roth and Bobko (2001) who found a negative correlation between race and college academic performance. Interestingly, although minority and female officers received lower performance appraisal ratings, they did not receive fewer commendations, have more disciplinary problems, or miss more days of work. Thus, it is difficult to determine if the lower performance ratings received by females and minorities is a true reflection of performance or a reflection of supervisor bias.

Older and more experienced officers received higher performance ratings and more commendations, had fewer injuries, and used force less often than their less experienced counterparts. However, they also had more discipline problems and engaged in less patrol activity than their less experienced peers.

Table 14.1: Meta-analysis results of the relationships between sex, race, age, and years on the force with criteria

Criterion/Demographic	K	N	*r*	95% Confidence Interval		ρ	90% Credibility Interval		Var	Q_w
				Lower	Upper		Lower	Upper		
Academy Performance										
Sex	13	3,876	-.10	-.17	-.03	-.14	-.42	.13	23%	56.91*
Race	10	3,950	-.29	-.36	-.21	-.41	-.60	-.21	48%	20.64*
Age	9	3,449	-.02	-.07	.03	-.03	-.17	.10	47%	19.23*
Performance Ratings										
Sex	18	2,451	-.07	-.12	-.02	-.12	-.32	.09	59%	30.52*
Race	15	2,609	-.18	-.25	-.12	-.27	-.50	-.04	49%	32.64*
Age	23	3,837	.05	-.01	.12	.09	-.29	.46	26%	89.88*
Years on the force	8	1,176	.23	.08	.38	.38	-.15	.91	22%	36.59*
Commendations										
Sex	2	232	-.01	-.13	.12	-.01	-.01	-.01	100%	0.05
Race	1	160	-.03							
Age	12	3,990	.09	.02	.15	.10	-.10	.31	26%	41.66*
Years on the force	3	721	.27	.11	.42	.35	.11	.59	35%	8.52*

K=number of studies, N=sample size, r = mean correlation, ρ = mean correlation corrected for range restriction, criterion unreliability, and predictor reliability, VAR = percentage of variance explained by sampling error and study artifacts, Q_w = the within group heterogeneity

Table 14.1: Meta-analysis results of the relationships between sex, race, age, and years on the force with criteria (cont.)

Criterion/Demographic	K	N	r	95% Confidence Interval		ρ	90% Credibility Interval		Var	Q_w
				Lower	Upper		Lower	Upper		
Activity										
Sex	2	600	-.04	-.30	.21	-.07	-.53	.38	10%	19.69*
Age	5	873	-.24	-.36	-.12	-.37	-.62	-.12	49%	10.30*
Years on the force	2	434	-.27	-.49	-.04	-.41	-.74	-.08	35%	5.76*
Discipline Problems										
Sex	7	4,299	.00	-.03	.03	.01	.01	.01	100%	5.28
Race	6	1,322	.03	-.02	.09	.06	-.22	.33	31%	19.54*
Age	14	6,459	.00	-.03	.03	.00	-.03	.04	92%	15.24
Years on the force	5	1,175	.16	.02	.31	.26	-.13	.66	20%	25.18*
Tenure										
Sex	3	1,706	.06	-.06	.17	.08	-.08	.24	50%	6.00*
Race	4	1,063	.00	-.11	.11	-.01	-.21	.20	31%	13.09*

K=number of studies, N=sample size, r = mean correlation, ρ = mean correlation corrected for range restriction, criterion unreliability, and predictor reliability, VAR = percentage of variance explained by sampling error and study artifacts, Q_w = the within group heterogeneity

Table 14.1: Meta-analysis results of the relationships between sex, race, age, and years on the force with criteria (cont.)

Criterion/Demographic	K	N	*r*	95% Confidence Interval		*ρ*	90% Credibility Interval		Var	Q_w
				Lower	Upper		Lower	Upper		
Absenteeism										
Race	2	320	.04	- .07	.15	.05	.05	.05	100%	0.13
Age	5	2,207	- .09	- .18	.01	- .12	- .22	- .01	80%	6.25
Accidents/Injuries										
Sex	1	299	- .12							
Race	2	320	.03	- .08	.14	.04	.04	.04	100%	0.13
Age	5	996	- .14	- .20	- .08	- .19	- .19	- .19	100%	2.05
Years on the force	1	299	- .02							
Use of Force										
Age	2	520	- .11	- .20	- .03	- .18	- .18	- .18	100%	0.33

K=number of studies, N=sample size, r = mean correlation, ρ = mean correlation corrected for range restriction, criterion unreliability, and predictor reliability, VAR = percentage of variance explained by sampling error and study artifacts, Q_w = the within group heterogeneity

Chapter References

*Aamodt, M. G., & Flink, W. (2001). Relationship between educational level and cadet performance in a police academy. *Applied H.R.M. Research, 6*(1), 75-76.

*Abraham, J. D., & Morrison, J. D. (2003). Relationship between the Performance Perspectives Inventory's conscientiousness scale and job performance of corporate security guards. *Applied H.R.M. Research, 8*(1), 45-48.

*Agyapong, O.A. (1988). *The effect of professionalism on police job performance: An empirical assessment.* Unpublished doctoral dissertation, Florida State University.

*Arvey, R. D., Landon, T. E., Nutting, S. M., & Maxwell, S. E. (1992). Development of physical ability tests for police officers: A construct validation approach. *Journal of Applied Psychology, 77*(6), 996-1009.

*Boyce, T. N. (1988). *Psychological screening for high-risk police specialization.* Unpublished doctoral dissertation, Georgia State University.

*Brewster, J., & Stoloff, M. (2003). Relationship between IQ and first-year performance as a police officer. *Applied H.R.M. Research*, *8*(1), 49-50.

*Buttolph, S. E. (1999). *Effect of college education on police behavior: Analysis of complaints and commendations.* Unpublished master's thesis, East Tennessee State University.

*Campa, E. E. (1993). *The relationship of reading comprehension and educational achievement levels to academy and field training performance of police cadets.* Unpublished doctoral dissertation, Texas A&M University.

*Champion, D. H. (1994). *A study of the relationship between critical thinking levels and job performance of police officers in a medium size police department in North Carolina.* Unpublished doctoral dissertation, North Carolina State University.

*Cohen, B., & Chaiken, J. M. (1973). *Police Background Characteristics and Performance.* Lexington, MA: Lexington Books.

*Copley, W. H. (1987). *Using education, academy, and field training scores to predict success in a Colorado police department.* Unpublished doctoral dissertation, Colorado State University.

*Dailey, J. D. (2002). *An investigation of police officer background and performance: An analytical study of the effect of age, time in service, prior military service, and educational level on commendations.* Unpublished doctoral dissertation, Sam Houston State University.

*Davis, R. D., & Rostow, C. D. (2003). Relationship between cognitive ability and background variables and disciplinary problems in law enforcement. *Applied H.R.M. Research, 8*(2), 77-80.

*Dorsey, R. R. (1994). *Higher education for police officers: An analysis of the relationships among higher education, belief systems, job performance, and cultural awareness.* Unpublished doctoral dissertation, University of Mississippi.

*Ford, J. K., & Kraiger, K. (1993). Police officer selection validation project: The multijurisdictional police officer examination. *Journal of Business and Psychology, 7*(4), 421-429.

*Geraghty, M. F. (1986). *The California Personality Inventory test as a predictor of law enforcement officer job performance.* Unpublished doctoral dissertation, Florida Institute of Technology

*Gonder, M. L. (1998). *Personality profiles of police officers: Differences in those that complete and fail to complete a police training academy.* Unpublished master's thesis, University of North Carolina-Charlotte.

*Griffin, G. R. (1980). *A study of relationships between levels of college education and police patrolmen's performance.* Saratoga, CA: Century Twenty One Publishing.

*Griffiths, R. F., & McDaniel, Q. P. (1993). Predictors of police assaults. *Journal of Police and Criminal Psychology, 9*(1), 5-9.

*Hamack, R. F. (1988). *Pre-academy placement in the Washington State Patrol: Factors associated with academy and job performance.* Unpublished master's thesis, Central Washington University.

*Hankey, R. O., Morman, R. R., Kennedy, P. K., & Heywood, H. L. (1965). TAV selection system and state traffic officer job performance, *Police*, March-April, 10-13.

*Hankey, R. O. (1968). *Personality correlates in a role of authority: The Police.* Unpublished doctoral dissertation, University of Southern California.

*Helrich, K. L. (1985). *The use of hardiness and othe stress-resistance resources to predict symptoms and performance in police academy trainees.* Unpublished doctoral dissertation, California School of Professional Psychology, San Diego.

*Hooper, M. K. (1988). *Relationship of college education to police officer job performance.* Unpublished doctoral dissertation, Claremont Graduate School.

*Hausknecht, J. P., Trevor, C. O., & Farr, J. L. (2002). Retaking ability tests in a selection setting: Implications for practice effects, training performance, and turnover. *Journal of Applied Psychology, 87*(2), 243-254.

*Kayode, O. (1973). *Predicting performance on the basis of social background characteristics: The case of the Philadelphia Police Department.* Unpublished doctoral dissertation, University of Pennsylvania

*Kedia, P. R. (1985). *Assessing the effect of college education on police job performance.* Unpublished doctoral dissertation, University of Southern Mississippi.

*Lester, D. (1979). Predictors of graduation from a police training academy. *Psychological Reports, 44*, 362.

*Lester, D. (1985). Graduation from a police training academy: Demographic correlates. *Psychological Reports, 57*, 542.

*Madden, B. L. (1990). *The police and higher education: A study of the relationship between higher education and police officer performance.* Unpublished master's thesis, University of Louisville.

*Matyas, G. S. (1980). *The relationship of MMPI and biographical data to police performance.* Unpublished doctoral dissertation, University of Missouri – Columbia.

*McConnell, W. A. (1967). *Relationship of personal history to success as a police patrolman.* Unpublished doctoral dissertation, Colorado State University.

* Mealia, R. M. (1990). *Background factors and police performance.* Unpublished doctoral dissertation, State University of New York, Albany.

*Morman, R. R., Hankey, R. O., Heywood, H. L., & Kennedy, P. K. (1965). Multiple relationships of TAV selection system predictors to state traffic officer performance, *Police*, July-August, 41-44.

*Morman, R. R., Hankey, R. O., Kennedy, P. K., & Heywood, H. L. (1965). Predicting state traffic officer performance with TAV selection system theoretical scoring keys, *Police*, May-June, 70-73.

*Morman, R. R., Hankey, R. O., Heywood, H. L., & Liddle, R. (1966). Predicting state traffic cadet academic performance from theoretical TAV selection system scores, *Police*, July-August, 54-58.

*Palmatier, J. J. (1996). *The big-five factors and hostility in the MMPI and IPI: Predictors of Michigan State Trooper job performance.* Unpublished doctoral dissertation, Michigan State University.

*Peterson, D. S. (2001). *The relationship between educational attainment and police performance.* Unpublished doctoral dissertation, Illinois State University.

* Pibulniyom, S. (1984). *A quantitative analysis of dynamic performance measurements of a southern police department.* Unpublished doctoral dissertation, University of Mississippi.

*Rose, J. E. (1995). *Consolidation of law enforcement basic training academies: An evaluation of pilot projects.* Unpublished doctoral dissertation, Northern Arizona University.

Roth, P. L., & Bobko, P. (2000). College grade point average as a personnel selection device: Ethnic group differences and potential adverse impact. *Journal of Applied Psychology, 85*(3), 399-406.

*Scarfo, S. J. (2002). Relationship between police academy performance and cadet level of education and cognitive ability. *Applied H.R.M. Research, 7*(1), 24.

*Schelling, M. J. (1993). *The use of the Law Enforcement Selection Inventory in the selection of communication officers: A concurrent validity study.* Unpublished master's thesis, Radford University.

*Schroeder, D. J. (1973). *A study of the validity of the entrance examination for the position of patrolman under the guidelines established the Equal Opportunity Employment Commission.* Unpublished master's thesis, John Jay College.

*Shaver, D. P. (1980). *A descriptive study of police officers in selected towns of northwest Arkansas.* Unpublished doctoral dissertation, University of Arkansas.

*Smith, S. M., & Aamodt, M. G. (1997). The relationship between education, experience, and police performance. *Journal of Police and Criminal Psychology, 12*(2), 7-14.

*Spurlin, O., & Swander, C. (1987). *Validity and fairness of the police officer written exam: Research finding.* Seattle, WA: Public Safety Civil Service Commission.

*Staff, T. G. (1992). *The utility of biographical data in predicting job performance: Implications for the selection of police officers.* Unpublished doctoral dissertation, University of Toledo.

*Stafford, A. R. (1983). *The relationship of job performance to personal characteristics of police patrol officers in selected Mississippi police departments.* Unpublished doctoral dissertation, University of Southern Mississippi.

*Tidwell, H. D. (1993). *The predictive value of biographical data: An analysis of using biodata to predict short tenure or unsuitability of police officers.* Unpublished doctoral dissertation, University of Texas - Arlington.

*Tomini, B. A. (1995). *The person-job fit: Implications of selecting police personnel on the basis of job dimensions, aptitudes and personality traits.* Unpublished doctoral dissertation, University of Windsor.

*Tompkins, L. P. (1986). *An evaluation of police academy training upon selected recruits and its relationship to job performance.* Unpublished master's thesis, Rollins College, FL.

* Wexler, N., & Sullivan, S. M. (1982). *Concurrent validation of a prototype selection test for entry-level police officer.* Trenton, NJ: New Jersey Department of Civil Service, Division of Examinations.

*Wilson, H. T. (1994). *Post-secondary education and the police officer: A study of its effect on the frequency of citizens' complaints.* Unpublished doctoral dissertation, Golden Gate University.

*Woods, D. A. (1991). *An analysis of peace officer licensing revocations in Texas.* Unpublished doctoral dissertation, Sam Houston State University.

*Wright, B. (1988). *Psychological evaluations as predictors of police recruit performance.* Unpublished doctoral dissertation, Florida State University

*Indicates study was used in the meta-analysis

Chapter 15
Summary and Recommendations

Chapters 3 through 14 contained many tables and lots of numbers. In this final chapter I will try to summarize what we now know about law enforcement selection and where we need to go to know more. To provide a quick review of the earlier chapters, Table 15.1 summarizes the validity and generalizability of the various predictors of law enforcement performance.

Education and Cognitive Ability

What We Know

- Educated, bright officers perform better in the academy, receive higher performance ratings, make more arrests, and issue more citations than do their less educated, less bright peers.
- Educated officers have fewer discipline problems, are absent less often, get into fewer automobile accidents, are less likely to be assaulted, and are less likely to use force.
- Education and cognitive ability are related (r = .29, ρ = .42) but the combination of education and cognitive ability is more valid than cognitive ability by itself (incremental validity).
- The performance differences between college-educated officers and their counterparts begin to show after a few years on the job (the honeymoon effect).
- Criminal justice majors do not perform better than other majors.

- Both cognitive ability tests and education requirements will have considerable adverse impact on minority hiring.

Where We Need to Go

- More research is needed to determine if there are performance differences between officers with associate's degrees and those with bachelor's degrees.
- Research is needed to determine the ideal cutoff score for cognitive ability. That is, at what point is an officer not smart enough and at what point do increases in cognitive ability not matter or perhaps even decrease performance?
- More research is needed on the relationships between cognitive ability and education and commendations, activity, absenteeism, injuries, use of force, and tenure.
- We know that increased education and cognitive ability are related to increased performance, but research is still needed to determine if the GPA earned in high school or college will help predict performance.

Background Variables

What We Know

- Other than receiving slightly more commendations ($r = .07$, $\rho = .10$), officers with military experience do not perform better or worse than their peers without military experience.
- Officers who had discipline problems at previous jobs have an increased likelihood of having discipline problems and performing poorly as a police officer ($r = -.21$, $\rho = -.34$).

Where We Need to Go

- On the basis of only two studies, the correlation between military experience and field training performance is promising (r = .15, ρ = .20). More research is needed here.

- Due to the small number of studies, more research is needed to investigate the relationship between prior military experience and the use of force and being assaulted.
- More research is needed to investigate the relationship between traffic citations received prior to hire and subsequent performance as a police officer. The research to date suggests that officers with many traffic citations will be poorer performers (r = - .12, ρ = - .20) and use more sick leave (r = .13, ρ = .17) but will also receive more commendations (r = .09, ρ = .13). However, these findings are based on only a few studies.

Vocational Interest Inventories

What We Know

- Scoring high on police interest scales is not related to supervisor ratings of police performance.
- On the basis of only five studies, high conventional interests (r = .14, ρ = .25) and low artistic interests (r = - .12, ρ = - .22) are related to higher supervisor ratings of performance.

Where We Need to Go

- Much more research is necessary in this area. More studies are especially needed comparing interest scores to such measures of performance as activity, commendations, and discipline problems.
- Because interest inventories are designed to predict satisfaction with careers, research is needed comparing interest scores with job satisfaction and tenure. That is, applicants scoring low on police

interest may perform at adequate levels, but due to low job or career satisfaction, do they quit the force or have higher levels of absenteeism?

Personality Inventories

What We Know

- Neither individual scale scores from measures of psychopathology nor individual psychopathology constructs (e.g., depression) are strong predictors of future law enforcement performance.
- The tolerance scale of the CPI (r = .20, ρ = .34) appears to be the single best personality predictor of law enforcement performance.
- The Good Cop/Bad Cop and Husemann Index are promising methods for using the MMPI, especially in terms of predicting discipline-related problems
- Psychologists' interpretations of test batteries may be useful predictors of law enforcement performance.
- The Big 5 factors of personality (openness, conscientiousness, extraversion, agreeableness, stability) were generally related to measures of law enforcement performance, but individual scales of the CPI had higher validities. Thus, the Big 5 may be too broad to tap the personality dimensions that will best predict law enforcement performance.
- The personality profile of the typical police officer indicates a psychologically healthy person.
- The "classic 4/9" spike (high scores on the psychopathic deviate and mania scales) found in police officers with the MMPI does not appear with the MMPI-2.

Where We Need to Go

- Cognitive ability and education are highly correlated with academy grades and research is needed to determine if personality can add incremental validity to these measures.
- Much more research is needed to determine if the Good Cop/Bad Cop Index and the Husemann Index will increase the usefulness of the MMPI-2.
- More research is clearly needed on the validity of psychologists' judgments about personality profiles.

Assessment Centers

What We Know

- On the basis of only four studies, assessment centers are valid (r = .17, ρ = .31) predictors of patrol performance.
- On the basis of only two studies, assessment centers are valid (r = .33, ρ = .57) predictors of supervisory performance.
- On the basis of non law-enforcement related research, assessment centers have excellent face validity but are expensive and time-consuming to administer.

Where We Need to Go

- With only six police-related studies involving assessment centers, more research is needed. Although assessment centers have been found to be good predictors of supervisory performance in non-law enforcement jobs, there is little research focusing on the validity of assessment centers for entry-level positions.
- Due to the expense of assessment centers, it is important to determine if assessment center scores will add incremental validity to other selection

methods and if the amount of incremental validity will justify the high cost.

Interviews

What We Know

- Panel interviews are valid predictors of academy grades (r = .12, ρ = .27) and supervisor ratings of performance (r = .09, ρ = .19).
- Interviews by clinical psychologists are not only poor predictors of law enforcement performance but actually reduce the validity of test scores.
- On the basis of non law-enforcement research, structured interviews using trained raters are better than unstructured interviews conducted by untrained raters.

Where We Need to Go

- In meta-analyses on non-law enforcement occupations, structured interviews have been shown to be one of the best predictors of employee performance. There is no reason to think that the same would not be true in law enforcement occupations. Thus, more research using structured interviews in law enforcement occupations would be a good idea.

Physical Agility

What We Know

- On the basis of only four studies, physical ability tests do not predict supervisor ratings of performance.
- On the basis of only two studies, physical ability tests predict supervisor ratings of physical fitness.

Where We Need to Go

- Much more research is needed investigating the relationship between physical ability tests and such criteria as supervisor ratings, patrol activity, injuries, and use of force.
- Research is needed to determine whether simulations (e.g., obstacle course, dummy drag) are more valid predictors of performance than individual physical ability tests (e.g., sit-ups, pull-ups).

Correlations among Criteria

What We Know

- Cadets who perform well in the academy will get higher supervisor ratings ($r = .22, \rho = .38$), make more arrests and write more citations ($r = .12, \rho = .40$), and have fewer disciplinary problems ($r = -.13, \rho = -.22$) than will cadets who do not perform as well in the academy.
- Discipline problems and commendations are not highly related but each is related to patrol activity. These findings support the notion that active officers will receive more complaints and more commendations than will their less-active counterparts.

Where We Need to Go

- Due to the small number of studies investigating relationships among criteria, more research is needed in this area.
- We know that active officers will receive more complaints and more commendations than less active officers. An interesting area for future research would

be to determine the "expected" ratio between complaints and commendations. That is, if we would expect the typical officer to receive five citizen complaints for every citizen commendation, officers who have higher or lower ratios can be evaluated accordingly.

Demographic Differences

What We Know

- Women have lower grades in the academy ($r = -.10$, $\rho = -.14$) and receive lower performance ratings ($r = -.07$, $\rho = -.12$) than do men.
- Minority officers ($r = -.29$, $\rho = -.41$) score lower in the academy and receive lower performance ratings ($r = -.18$, $\rho = -.27$) than do nonminorities.
- On the basis of a limited number of studies, there do not appear to be any sex or race differences on objective measures of performance.
- Older officers receive more commendations ($r = .09$, $\rho = .10$) and are injured less often ($r = -.14$, $\rho = -.19$) than their younger counterparts but also engage in less patrol activity ($r = -.24$, $\rho = -.37$).

Where We Need to Go

- Much more research is needed in this area. It is important to determine if the sex and race differences in supervisor ratings of performance are due to actual differences in performance or due to discrimination in the rating process.

Table 15.1 Answers to Basic Meta-Analysis Questions

Predictor/Criterion	K	Meta-Analysis Question		
		Significant Predictor?	True Validity (ρ)	Can results be generalized?
Cognitive ability				
Academy grades	61	Yes (r = .41)	.62	Yes
Performance ratings	61	Yes (r = .16)	.27	Yes
Commendations	7	No (r = -.01)	- .02	Yes
Activity	6	Yes (r = .19)	.33	Yes
Discipline	13	No (r = -.06)	- .11	No
Education				
Academy grades	32	Yes (r = .26)	.38	Yes
Performance ratings	54	Yes (r = .17)	.28	Yes
Activity	17	Yes (r = .05)	.09	Yes
Commendations	24	No (r = - .03)	- .04	No
Discipline problems	50	Yes (r = -. 07)	- .12	Yes
Absenteeism	18	Yes (r = - .10)	- .14	Yes
Injuries	10	Yes (r = -. 06)	- .08	No
Use of force	10	Yes (r = - .07)	- .12	Yes
Military Experience				
Academy grades	9	No (r = .02)	.04	Yes
Performance ratings	16	No (r = - .03)	- .05	No
Commendations	8	Yes (r = .07)	.10	Yes
Discipline problems	14	No (r = - .02)	- .04	Yes
Police Interest				
Performance ratings	8	No (r = -.03)	- .06	Yes
Interviews				
Performance ratings	8	Yes (r = .09)	.19	Yes

The Ideal Selection Battery

After I presented the preliminary results of these meta-analyses at the annual meeting of the Society for Police and Criminal Psychology in October, 2003, an audience member asked how the results could be used in recommending the "ideal" selection battery. On the basis of what we know from published research, dissertations, and theses, the "ideal" selection battery would contain the following components:

1. Only candidates with at least an associate's degree, and perhaps a bachelor's degree, should be considered for employment. Though this recommendation is justified by the validity evidence, it is important to note that some locations may find that education requirements cause too great a reduction in the applicant pool.
2. Applicants should complete a background questionnaire and those with a *pattern* of disciplinary problems (e.g., fired from work, expelled from school) or legal problems (e.g., arrests, traffic citations) should be eliminated. Answers to the background questionnaire can be confirmed at a later time during a polygraph examination and a background investigation.
3. A cognitive ability test should be used to ensure that applicants have the ability to acquire knowledge during training and apply that knowledge on the job. Tests specifically designed for police selection or tests of reading ability would be best.
4. The tolerance scale of the CPI should be used. This recommendation is not to imply that others tests or scales are not useful, but at this time, the tolerance scale has the most research support.
5. A structured interview should be used to tap an applicant's social skills, communication skills,

and problem-solving ability. All interview questions should be job-related (based on a job analysis), every applicant should be asked the same questions, and a standard scoring system should be used for all questions.

6. After a conditional offer of employment has been made, the applicant should undergo medical and psychological exams. The psychological exam should consist of a test of psychopathology (e.g., MMPI-2, IPI, PAI). Although the meta-analyses did not support measures of psychopathology being predictive of law enforcement performance, from a liability perspective, it would seem reasonable to screen out applicants with clear indications of work-related psychopathology. The physician conducting the medical exam should be given a thorough job description and instructed to determine if there is any medical reason why the applicant cannot perform the essential functions of the job.
7. A background investigation and polygraph examination should be used to verify information provided by the applicant in the background questionnaire.

Comparison with Other Meta-Analyses

The meta-analyses discussed in this book included only studies using law enforcement samples. As shown in Table 15.2, the results are at times different from meta-analyses using non-law enforcement samples. Perhaps the greatest differences are in cognitive ability and education. Education is a better predictor of law enforcement performance than of performance in other occupations. On the contrary, cognitive ability is a better predictor of performance in other occupations than it is in law enforcement. In terms of personality, openness to experience appears to be a better predictor of law enforcement performance than performance in other occupations.

Table 15.2 Comparison of this meta-analysis to other meta-analyses

Criterion/Predictor	Law Enforcement		All Occupations		
	r	*ρ*	*r*	*ρ*	Authors
Performance Ratings					
Cognitive ability	.16	.27		.51	Hunter & Hunter (1984)
Education	.17	.28		.10	Hunter & Hunter (1984)
Interest	- .04	- .06		.10	Hunter & Hunter (1984)
Interviews (Unstructured)	.09	.19	.11	.20	Huffcutt & Arthur (1994)
Assessment centers	.17	.31	.25	.36	Gaugler et al. (1987)
Personality					
Openness	.08	.14	.02	.04	Barrick & Mount (1991)
Conscientiousness	.12	.22	.15	.26	Barrick & Mount (1991)
Extraversion	.05	.09	.08	.14	Barrick & Mount (1991)
Agreeableness	.07	.13	.05	.09	Barrick & Mount (1991)
Stability	.09	.17	.05	.09	Barrick & Mount (1991)
Training					
Cognitive ability	.41	.62		.56	Hunter & Hunter (1984)
Education	.26	.38		.20	Hunter & Hunter (1984)
Interviews	.12	.27			
Assessment centers	.22	.37	.30	.35	Gaugler et al. (1987)
Personality					
Openness	.22	.34	.14	.25	Barrick & Mount (1991)
Conscientiousness	.13	.21	.13	.23	Barrick & Mount (1991)
Extraversion	.12	.19	.15	.26	Barrick & Mount (1991)
Agreeableness	.04	.06	.06	.10	Barrick & Mount (1991)
Stability	.11	.17	.04	.07	Barrick & Mount (1991)

A Call for Further Research

At the beginning of this book I stated that my intended goal was to review all of the current research on law-enforcement selection. I think that I have done this and that this final chapter is an accurate summary of our current knowledge. However, as was stated many times during this chapter, more research is needed.

Perhaps the reason for the small number of studies in some areas is that researchers think that because "someone already did that" a similar study would not be useful. Nothing could be further from the truth. As can be seen throughout this book, the more studies that are conducted on a topic, the better picture we have of "the truth." Some might respond that it is difficult to publish a study on a topic that has been published five times before, and I won't argue with that. That is why one journal, *Applied H.R.M. Research*, encourages researchers to publish 1 to 4 page Validity Studies. These validity studies are short summaries similar to those found in the appendix to this book. Information about submitting articles to this journal can be found at www.radford.edu/~applyhrm.

When publishing or presenting the results of a validation study, it is essential that future researchers provide a complete set of correlations. Publishing only the significant correlations makes it difficult to determine the validity of scores from tests and other employee selection techniques. It is also important that future researchers include information on the reliability of the predictors, the reliability of the performance criteria, and information that can be used to assess the degree of range restriction in the sample.

Did I Miss a Study?

My goal was to include all relevant studies in the meta-analyses conducted for this book. If you are aware of a study that was not included, please feel free to email me or send me a copy of the study so that it can be included in future editions. My contact information is:

Dr. Mike Aamodt
Department of Psychology
Radford University
Radford, VA 24142-6946
(540) 831-5513
maamodt@radford.edu

Chapter References

Barrick, M. R., & Mount, M. K. (1991). The big five personality dimensions and job performance: A meta-analysis. *Personnel Psychology, 44*(1), 1-26.

Gaugler, B. B., Rosenthal, D. B., Thornton, G. C., & Bentson, C. (1987). Meta-analysis of assessment center validity. *Journal of Applied Psychology, 72*(3), 493-511.

Huffcutt, A. I., & Arthur, W. (1994). Hunter and Hunter (1984) revisited: Interview validity for entry-level jobs. *Journal of Applied Psychology, 79*(2), 184-190.

Hunter, J. E., & Hunter, R. F. (1984). Validity and utility of alternative predictors of job performance. *Psychological Bulletin, 96*(1), 72-98.

Subject Index

Name Index

Printed in the United States
39245LVS00002B/29